Emerging Technologies in the Construction Industry

Matthew N. O. Sadiku, Ph.D., P.E.

Regents Professor Emeritus and IEEE Life Fellow
Prairie View A&M University
Prairie View, TX 77446
Email: sadiku@ieee.org
Web: www.matthew-sadiku.com

BOOK FILMS MEDIA

Date published: March 07, 2025

Matthew N. O. Sadiku
sadiku@ieee.org
www.matthew-sadiku.com

Paperback: 979-8-9922654-9-1
eBook: 979-8-9924816-0-0
Hardcover: 979-8-9915027-5-7

BookFilmsMedia
2780 South Jones Blvd Suite 200 - 4007
Las Vegas, NV 89146 United States
+1 725 - 238 - 6534

Photo on Cover: https://parametric-architecture.com/how-3d-printing-technology-can-be-used-to-tackle-the-global-housing-crisis/

DEDICATED TO MY FRIENDS:

CHRISTOPHER O. FALADE
CHRISTIANAH I. FALADE

BRIEF TABLE OF CONTENTS

PREFACE

The construction industry, also known as the building industry, is one of the oldest businesses that humans developed. It is one of the largest and most labor-intensive industries in the world. From building neighborhood homes to mile-long suspension bridges, construction is the foundation of the growth and essential to the development and advancement of any nation.

Technology changes our lives constantly. It has come a long way when we refer to the construction industry. Technology has enabled construction workers and managers to access real-time data, allowing them to make informed decisions on the spot. Modern construction managers now find themselves at the intersection of technology and construction. Construction has been well known for its labor-intensive nature, hazardous working conditions, and requirement for exact coordination between personnel, equipment, and supplies.

The construction industry, a cornerstone of global infrastructure development, is undergoing a significant transformation, driven by the rapid adoption of digital technologies and using technologies to tackle issues like productivity, safety, labor shortages, and collaboration. The industry is experiencing a wave of emerging technologies that promise to transform the way things are built. At the heart of this transformation lie some emerging technologies. If construction companies want to remain relevant, they must invest in new and emerging technologies.

This book examines emerging technologies and their impact on the construction industry. It is organized into twelve chapters that summarize emerging technologies and their applications in construction.

Chapter 1 - Introduction:

This chapter addresses the implementation of emerging technologies in the construction industry and serves as an introduction to the entire book. It introduces the readers to some emerging technologies in the construction industry, including artificial intelligence, robotics, drones, cloud computing, Internet of things, blockchain, 3D printing, wearables, augmented reality, cybersecurity, and building information modeling. Understanding these new technologies allows industry players to stay resilient in the face of an increasingly challenging landscape.

Chapter 2 - Artificial Intelligence:

This chapter aims to unravel artificial intelligence (AI) applications in modern construction industry. Artificial intelligence has become integral to construction processes, offering efficiency, precision, and innovation. From optimizing the design process to reducing energy consumption, AI has significant implications for the construction industry, making it more efficient and sustainable. Recent research indicates that applications of AI technologies are increasingly permeating the construction industry in all areas of architectural design, engineering design, and construction services.

Chapter 3 - Robotics:

In this chapter, we will explore how robotics is making sweeping changes in the construction industry. Robotics is the branch of engineering that integrates computer science with mechanical and electrical engineering. It is now finding its way into the construction industry. Robots usually speed up production and eliminate the cost of labor. Imagine a team of robots working along with laborers on a construction site to transform the landscape. The construction industry is undergoing a robotic overhaul. Robotics is poised to redefine the construction landscape, making it more efficient, precise, and sustainable.

Chapter 4 - Drones:

This chapter explores various ways of using drones in the construction industry. Drone photos, videos, and imagery are used to scope out projects, track building progress, and provide real-time updates. With their real-time data recording and unique aerial advantage, drones can improve efficiency, cut costs, and streamline workflow. Drones offer immense possibilities and applications in the world of construction, offering benefits ranging from on-site safety to remote monitoring. While the impact of drones in the construction industry is evolving, they are already revolutionizing the sector and changing the way construction projects are handled.

Chapter 5 - Internet of Things:

In this chapter, we will explore how the construction industry is embracing the Internet of things (IoT) as a transformative tool. The Internet of things is a system of interrelated smart devices that are fitted with sensors, software, and other technologies that facilitates communication. IoT technology in construction is improving safety, efficiency, and project management, revolutionizing construction operations and driving innovation. The construction industry is in the research stage of the IoT adoption journey. IoT technology has proven to be a magnificent innovation in the construction industry by improving its working condition, reducing waste, and increasing its efficiency.

Chapter 6 - Cloud Computing:

This chapter investigates the extent to which cloud computing has been used in the construction industry. Cloud computing (CC) is a model for enabling convenient, on-demand network access to a shared pool of configurable computing resources (e.g., networks, servers, storage, applications, and services) that can be rapidly provisioned. It has emerged as a transformative force in the construction industry, offering countless benefits such as cost savings, improved collaboration, and enhanced project management. It will result in emergence of new business models which will enable construction companies to do business in a new way.

Chapter 7 - Blockchain:

This chapter explores how blockchain technology could potentially change the construction industry for the better. Blockchain is a distributed database that allows for secure, transparent, and tamper-proof transactions. It is a system that tracks transactions across a peer-to-peer network. Its main defining feature is decentralization. Its adoption is a trend that will help the construction industry become more efficient overall. Construction businesses may improve supply chain efficiency, cut down on delays, and minimize risks by embracing the transparency and traceability offered by blockchain technology.

Chapter 8 - 3D Printing:

In this chapter, we examine the use of 3D printing technology in the construction industry. 3D printing, also known as additive manufacturing, is a process of creating a physical object from a digital model. The construction industry is one of the most exciting applications of 3D printing technology because structures can be built using a fraction of the time, labor, and material compared to traditional structures. The construction industry is changing due to 3D printing, making it cheaper and faster to construct buildings and other structures by automating building processes. The technology has the potential to revolutionize the construction of buildings and structures, saving time, materials, labor, and costs, while also enhancing the sustainability and environmental impact of construction.

Chapter 9 - Wearables:

This chapter discusses the uses of wearable technologies in the construction industry. As the construction industry prioritizes worker safety, wearable technology has emerged as a promising solution. Wearable technology is ushering in a new era of construction. Wearable construction technology refers to any device that construction workers wear on their bodies, to improve safety and reduce injuries. The future of safe construction involves user-friendly and powerful wearable technology, including smart watches, boots, helmets, glasses and body wear.

Chapter 10 - Augmented Reality:

In this chapter, we will discuss how augmented reality is used in construction. Augmented reality (AR) is an immersive technology that overlays digital information, such as images, videos, or

3D models, onto the real-world environment in real-time. The construction industry has adopted AR technology to improve the design, planning, and construction processes. AR technology allows architects and designers to visualize 3D models of their designs in a real-world environment. It has the potential to revolutionize the construction industry. It benefits the entire construction team: engineers, designers, architects, project managers, and service providers.

Chapter 11 - Building Information Modeling:

The aim of this chapter is to examine how building information modeling (BIM) is used in the construction industry. BIM is one of the most promising developments in the architecture, engineering, and construction (AEC) industries. It is a digital representation of physical and functional characteristics of a building or other physical asset and facility. BIM technology is a revolutionary approach to the design, construction, and management of buildings and infrastructure. It allows construction managers and architectural professionals to communicate easily and build more efficiently, while giving clients a higher quality finished product. It has become the standard for modern construction projects.

Chapter 12 - Cybersecurity:

This chapter discusses the role of cybersecurity in the construction industry. The construction industry is now confronting a new frontier of threats emanating from the digital realm. Today, the built environment is designed, constructed, and managed using digital technology, making it increasingly exposed to cyber security risks. The construction industry is at a crossroads of innovation and vulnerability. The industry is increasingly becoming a prime target for cyberattacks. It is confronted with various cybersecurity challenges as it continues to adopt digital technologies and innovative solutions. As the digital transformation accelerates, construction companies must recognize that cybersecurity is critical to their bottom line and business continuity.

This is a comprehensive introductory text on the issues, ideas, theories, and applications of emerging technologies in the construction industry. It provides an overview of each technology in simple terms so that beginners can understand it. It is a must-read book for anyone who wants to learn about modern construction, which have become vital in many areas of life.

— M. N. O. Sadiku and C. O. Falade

ABOUT THE AUTHOR

Matthew N. O. Sadiku received his B. Sc. degree in 1978 from Ahmadu Bello University, Zaria, Nigeria and his M.Sc. and Ph.D. degrees from Tennessee Technological University, Cookeville, TN in 1982 and 1984 respectively. From 1984 to 1988, he was an assistant professor at Florida Atlantic University, Boca Raton, FL, where he did graduate work in computer science. In total, he received seven college degrees. From 1988 to 2000, he was at Temple University, Philadelphia, PA, where he became a full professor. From 2000 to 2002, he was with Lucent/Avaya, Holmdel, NJ as a system engineer and with Boeing Satellite Systems, Los Angeles, CA as a senior scientist. He is presently a Regents professor emeritus of electrical and computer engineering at Prairie View A&M University, Prairie View, TX.

He is the author of over 1,350 professional papers and over 140 books including "Elements of Electromagnetics" (Oxford University Press, 7th ed., 2018), "Fundamentals of Electric Circuits" (McGraw-Hill, 7th ed., 2020, with C. Alexander), "Computational Electromagnetics with MATLAB" (CRC Press, 4th ed., 2019), "Principles of Modern Communication Systems" (Cambridge University Press, 2017, with S. O. Agbo), and "Emerging Internet-based Technologies" (CRC Press, 2019). In addition to the engineering books, he has written Christian books including "Secrets of Successful Marriages," "How to Discover God's Will for Your Life," and commentaries on all the books of the New Testament Bible. Some of his books have been translated into ten languages: French, Korean, Chinese (and Chinese Long Form in Taiwan), Italian, Portuguese, Spanish, German, Dutch, Polish, and Russian.

He was the recipient of the 2000 McGraw-Hill/Jacob Millman Award for outstanding contributions in the field of electrical engineering. He was also the recipient of Regents Professor award for 2012-2013 by the Texas A&M University System. He is a registered professional engineer and a life fellow of the Institute of Electrical and Electronics Engineers (IEEE) "for contributions to computational electromagnetics and engineering education." He was the IEEE Region 2 Student Activities Committee Chairman. He was an associate editor for IEEE Transactions on Education. He is also a member of Association for Computing Machinery (ACM). His current research interests are in the areas of computational electromagnetic, computer science/networks, engineering education, and marriage counseling. His works can be found in his autobiography, "My Life and Work" (Author's Tranquility Press, 2024) or his website: www.matthew-sadiku.com. He currently resides with his wife Janet in Westlake, Florida. He can be reached via email at sadiku@ieee.org

TABLE OF CONTENTS

CHAPTER 1
INTRODUCTION

"Each of us is carving a stone, erecting a column, or cutting a piece of stained glass in the construction of something much bigger than ourselves."

—Adrienne Clarkson

1.1 INTRODUCTION

Technology changes our lives constantly. Traditionally, construction managers relied on manual methods and paper-based documentation to perform key duties such as overseeing labor, costs, and schedules. But over the decades, technology like power tools and heavy-duty equipment emerged, improving productivity and safety. Technology in construction has truly come a long way for an industry that has the unfortunate reputation of being "slow to adopt." Modern construction managers now find themselves at the intersection of technology and construction. Larger construction enterprises have been experimenting with the use of new technologies for a while. There has never been a better time for technology adoption in construction. Figure 1.1 shows a building under construction [1].

Figure 1.1 A building under construction [1].

The field of construction has expanded at an astounding pace in the past decade. The digitization of the construction industry is perhaps most clearly seen in the proliferation of cloud-based data analytics and project management platforms tailored to the sector's needs. New technologies have continually revolutionized the way construction projects are managed and delivered. Technology has enabled construction workers and managers to access real-time data, allowing them to make informed decisions on the spot. Figure 1.2 shows some construction workers [2].

Figure 1.2 Some construction workers [2].

In spite being a $12 trillion giant, the architecture, engineering, and construction (AEC) industry has historically been slow to embrace new technologies, largely due to a lack of understanding and implementation support. Today, technology has been a driving force in the construction industry. The industry has continually evolved, seeking new methods to enhance efficiency. It is experiencing a wave of emerging technologies that promise to transform the way things are built. These technologies include artificial intelligence, machine language, robotics, Building Information Modelling (BIM), cloud-based project management, virtual reality, 3D printing, Internet of things, and blockchain [3].

This chapter addresses the implementation of emerging technologies in the construction industry and serves as an introduction to the entire book. It begins with describing what emerging technologies are all about. It briefly presents some emerging technologies in the construction industry. It highlights the benefits and challenges of the emerging technologies in the construction industry. It concludes with comments.

1.2 WHAT ARE EMERGING TECHNOLOGIES?

Technology may be regarded as a collection of systems designed to perform some function. It can help alleviate some of the challenges facing business today. Emerging technology is a term generally used to describe new technology. The term often refers to technologies currently developing or expected to be available within the next five to ten years. Any imminent, but not fully realized, technological innovations will have some impact on the status quo.

Emerging technologies are shaping our societies. They continue to affect the way we live, work, and interact with one another. Emerging technology (ET) lacks a consensus on what classifies them as "emergent." It is a relative term because one may see a technology as emerging and others may not see it the same way. It is a term that is often used to describe a new technology. A technology is still emerging if it is not yet a "must-have" [4]. An emerging technology is the one that holds the promise of creating a new economic engine and is trans-industrial. ET is used in different areas such as media, healthcare, business, science, education, or defense.

The characteristics of emerging technologies include the following [5]:

- *Novelty:* Emerging technologies are typically new or novel, meaning they have yet to be widely adopted or used. They often represent a significant departure from existing technologies or processes.
- *Potential for Disruption:* Emerging technologies have the potential to disrupt existing markets, industries, or ways of doing things. They may also displace existing businesses or industries.
- *Uncertainty:* Because emerging technologies are still in the early stages of development, there is often a high uncertainty surrounding their future potential and impact. It can be challenging to predict how they will evolve.
- *Rapid Change:* Emerging technologies often evolve rapidly, with new developments and innovations emerging frequently. It can make keeping up with the latest trends and advancements challenging.
- *Interdisciplinary:* Emerging technologies often involve multiple disciplines or fields of study, such as computer science, engineering, and biology. They may require collaboration across different fields and industries to develop their potential fully.

Emerging technologies are worth investigating. They are responsible for developing new products or devices. As emerging technologies continue to evolve, engineering is poised for a transformative future. Emerging technologies have driven innovation and progress in today's rapidly evolving digital landscape. The collective impact of emerging technologies such as artificial intelligence, machine learning, big data, and the Internet of things is undeniably transformative.

1.3 EMERGING TECHNOLOGIES IN CONSTRUCTION

The emergence of artificial intelligence, machine language, robotics, Building Information Modelling (BIM), cloud-based project management, virtual reality, and blockchain has transformed construction industry, allowing for more precise planning, increased productivity, and better resource management. We now consider the following emerging technologies in the construction industry [6]:

1. *Artificial Intelligence:* AI has provided benefits to construction projects through increased safety, improving workflows, and getting jobs done faster and better. It is employed for project scheduling and resource management, significantly improving efficiency. Machine Learning is a method of data analysis based on the idea that systems can learn from data, identify patterns and make decisions with minimal human intervention. Machine learning is helping construction pros optimize their workflows and aid decision-making. AI is increasingly shaping the future of the construction industry by enhancing safety, streamlining workflows, and accelerating project completion. Construction Management Association of America (CMAA) proclaimed AI has the potential to increase industry profits by 71% by 2035. Generative AI is transforming the construction industry by streamlining design processes and enabling smarter, data-driven decision-making. It can streamline processes, automate repetitive tasks, and optimize resource allocation. Architects and engineers leverage this technology to produce multiple design iterations, optimize material usage, and simulate environmental impacts, delivering more efficient and sustainable results.

2. *Robotic Automation:* The adoption of robotics and automation in the construction industry is transforming how construction companies execute complex construction projects. Robotics is transforming construction sites, with robots being used for tasks such as bricklaying, material handling, demolition, and even 3D printing of entire structures. Robotics in construction is revolutionizing the way tasks like bricklaying, demolition, and

excavation are conducted. These robotic tools not only reduce the necessity for manual labor, thus decreasing potential injuries and worker fatigue but also boost efficiency and enhance safety on-site. By automating routine and repetitive tasks, the construction industry can mitigate the effects of labor shortages. The integration of robotics in the construction industry faces challenges such as high initial costs and a skills gap within the workforce. Robotics and other automated technologies also enhance productivity, allowing skilled workers to focus on more complex aspects of construction. The continued adoption of robotics in construction is another significant development that will shape the industry's future. Figure 1.3 shows a typical workflow automation [7].

Figure 1.3 A typical workflow automation [7].

3. *Drones:* Many construction sites are heavily relying on the use of drones, also known as, unmanned aerial vehicles (UAV). Drones are the most widely used emerging construction technology. They have become a cornerstone technology in the construction industry, offering significant advantages for surveying and inspecting sites. They give contractors an overhead and close-up view with images and videoS, and collecting data with other technologies like thermal sensors and GPS tools. They are reshaping construction technology by providing advanced tools for site management, safety, and data collection. They revolutionize site

surveying by capturing aerial images and generating detailed maps. Not only do they offer a bird's-eye view for project monitoring and surveying, but construction managers can use drones to conduct site inspections, monitor progress, and capture high-resolution images for detailed analysis. One of the key benefits of drones is their ability to access and inspect hard-to-reach areas such as bridges and the exteriors of tall buildings safely. Utilizing drones and sensors for continuous surveillance improves response times to potential safety hazards, ensuring constant oversight. Drones not only enhance the safety of inspections by keeping workers off dangerous heights but also bring efficiency to data collection across expansive or difficult terrain. Figure 1.4 illustrates a construction drone [8].

Figure 1.4 A construction drone [8].

4. *Internet of Things (IoT):* The Internet of Things is a larger concept that prioritizes interconnected systems and has brought about a new era of connectivity on construction sites. Sensors and smart devices are embedded in construction equipment, tools, and even building materials, providing real-time data on performance, usage, and safety. In the construction industry, the IoT can allow construction machinery, equipment, materials, structures, and more to talk to a central data platform. We must not forget about the impact of IoT on the end-user experience. Imagine a world

where the buildings we construct are not just functional, but also intelligent, creating a truly personalized living or working environment.

5. *3D Printing:* 3D printing in the construction industry has seen remarkable advancements, particularly in 3D concrete printing. As a construction technology, it has the potential to change material sourcing. Materials such as concrete and composite materials can be used in these printers, enhancing design flexibility and the structural integrity of buildings. Although 3D printing currently faces challenges in mass production due to the high cost of large-scale printers, it offers substantial benefits, including faster construction timelines, cost savings, and the ability to create complex, customized architectural designs. 3D printed construction materials are often cost effective due to the decreased need for labor and shipping costs associated with traditional construction methods. 3D printing makes it possible to print materials right on site, reducing waste and further saving on transportation and storage costs. Figure 1.5 shows 3D printing in construction [9].

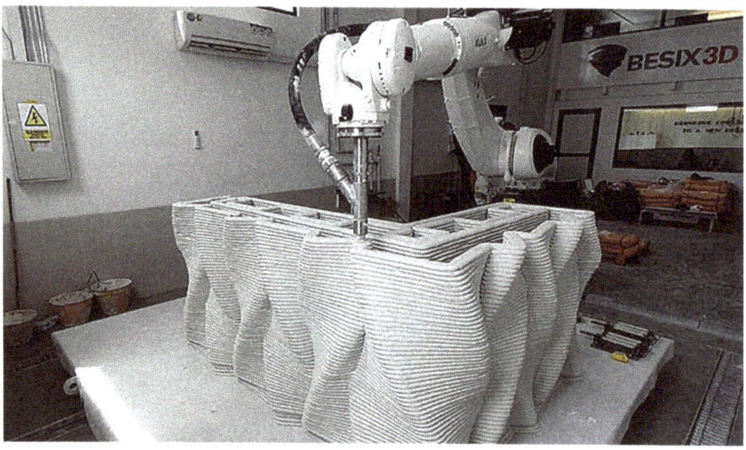

Figure 1.5 3D printing in construction [9].

6. *Blockchain:* Blockchain technology is rapidly becoming a transformative force in the AEC industry. Blockchain serves as a digital ledger, composed of linked data "blocks" that record every transaction and completed milestone in a project. This setup not only enhances cost management and procurement strategies

but also introduces a high level of accuracy and security. As we look ahead, the decentralized nature of blockchain makes it a robust, adaptable tool for projects of any size. The Architecture, Engineering, and Construction (AEC) industry is poised to enhance cost management and procurement strategies through the adoption of blockchain technology. Figure 1.6 depicts a construction worker using blockchain [9].

Figure 1.6 A construction worker using blockchain [9].

7. *Wearables:* Wearables are a construction technology that are not only making a positive impact on safety, but also productivity. They are now being built into all kinds of construction apparel such as safety vests, gloves, hard hats, and work boots. Beyond safety, wearables can also be embedded with other technology such as cameras and GPS modules. Wearables in the construction industry go beyond traditional personal protective equipment like hard hats, gloves, and work boots, enhancing safety and productivity significantly. Smart clothing, or e-textiles, further extends the capabilities of wearables in construction.

8. *Virtual Reality:* Extended Reality (XR) serves as the encompassing term for a spectrum of captivating and interactive technologies, including Virtual Reality (VR), Augmented Reality (AR), and Mixed Reality (MR). Virtual and augmented reality remain transformative technologies in the construction industry, offering advanced visualizations and enhancing collaboration.

VR/AR enhances teamwork by enabling real-time communication and collaboration across dispersed teams. Virtual reality (VR) is transforming traditional construction methods, allowing for more accurate spatial planning and project execution, VR technology is used in safety training programs to simulate realistic scenarios. This allows workers to gain experience and practice handling dangerous situations in a controlled, risk-free environment. VR technology allows stakeholders to visualize and interact with project designs before construction begins, offering a powerful tool for enhancing understanding and communication between clients, designers, and builders. This technology plays a pivotal role in revolutionizing Architectural, Engineering, and Construction (AEC) workflows. Architects and construction workers now have access to virtual reality technology that enables them to accurately test out designs, plan projects and create virtual 3D models of their build specifications. By implementing the virtual reality in the process, the process of visualizing of 2-D drawings or mock-up construction is eliminated. Figure 1.7 shows the use of VR in construction planning [8].

Figure 1.7 Use of VR in construction planning [8].

9. *Cybersecurity:* This has become a critical concern as reliance on digital tools increases. Ensuring the security of sensitive data is paramount as more aspects of construction operations go digital. In an industry where sensitive data and critical infrastructure collide, the need for robust security measures is paramount. The

construction industry is stepping up its game, investing in advanced encryption, threat detection, and incident response protocols.

10. *Building Information Modelling (BIM):* Another hot construction technology trend is BIM which promises to bring more accuracy to the building process. A construction team comprises many members, each one responsible for several aspects of the building process that will eventually bind together upon the completion of a project. BIM pairs architects with engineers and construction professionals. It is an intelligent process that allows architects, contractors, and engineers to collaborate on the construction, design, and operation of a building. It has the ability to depict project development in an open and highly collaborative environment. The use of BIM provides space for better collaboration because each person and expertise area can add their piece to the same model, instead of broken out onto multiple versions of a 2D paper drawing.

BIM streamlines information sharing, adding value to the data and enabling its reuse. As a result, teams can work more effectively and avoid errors or redundancies. It also helps with problem solving in the design and planning stages of a project, by automating clash detection and providing a more complete picture of the project. While BIM is primarily used for buildings, a similar technology, bridge information modeling, provides the same functionality for bridge builders. For example, BIM was extensively used in the creation of the Shanghai Tower, an internationally recognized architectural project, shown in Figure 1.8 [9].

Figure 1.8 BIM was extensively used in the creation of the Shanghai Tower [9].

11. *Cloud Technology:* Cloud operating systems have already been in use in many industries. Today, with mobile devices leveraging cloud technology (anywhere, anytime), you can look forward to storing massive amounts of information, and sharing it instantly. Cloud solutions can not only facilitate storing large amounts of data but also in making it accessible for project managers, keeping it secure in a remote location. With construction projects featured with high fragmentation, and different teams having little or no synchronization, you can expect cloud-based collaboration solutions, addressing the gap efficiently.

1.4 BENEFITS

Technological advancements are helping to overcome traditional challenges such as project delays and costs, labor shortages, and safety concerns. Emerging technologies are helping construction companies eliminate long-standing inefficacies and low productivity. The adoption of emerging technologies is significantly transforming the construction industry, enhancing efficiency, productivity, and safety throughout the construction process. Other benefits in the implementation of the emerging technologies include the following [10]:

- *Real-Time Data and Analytics:* Technology can provide real-time data on workforce productivity and project progress.

Construction companies can use this data to identify areas for improvement, optimize labor allocation, and make data-driven decisions about staffing needs.

- *Increased Safety:* It goes without saying that safety is paramount in construction and it always will be. Construction technology has evolved significantly over time, driven by increasing demands for efficiency and safety. It can not only mitigate risks but also foster a culture of safety that spreads through all levels of a construction project. Being able to constantly monitor worker safety will alleviate a lot of pressure operators face in the construction industry, drastically reducing both the fatality and injury rates.
- *Automation:* Technology can streamline and automate payroll processes, reducing errors and saving time. For example, cloud-based payroll systems can automate tasks like calculating overtime and deductions, ensuring workers are paid accurately and on time.
- *Increased Productivity:* Cloud-based construction management software automates various routine tasks, leading to substantial time savings and thus increased productivity.
- *Attracting New Talent:* As construction becomes more reliant on technology, the demand for workers with digital skills will increase. The integration of advanced technologies such as 3D printing, AI, and robotics into construction roles makes the field more appealing to a new generation of workers. These technologies offer a modern and engaging work environment, attracting individuals interested in tech-driven careers.
- *Efficiency:* Modern project management software has become a cornerstone of efficient operations for construction companies. Emerging technologies AI and 3D printing are revolutionizing the construction industry. These technologies automate tedious tasks, significantly increasing efficiency, reducing the need for a large workforce, and allowing

technicians to concentrate on specialized tasks.

- *Self-Healing Concrete:* Self-healing concrete is a miraculous material that shows promise for bridge- and roadwork. This type of concrete is created by adding special fibers or capsules containing adhesive liquids into the concrete mix. It has reparative properties and can seal cracks autogenously or autonomously.

- *Enhanced Collaboration:* These platforms facilitate improved collaboration among teams by providing a centralized location for all project-related documents, drawings, and schedules.

- *Costs:* AI can also be used to predict costs and identify areas where money can be saved, which is great for budget management. For example, AI-enabled systems can help assess material tracking and project build size.

1.5 CHALLENGES

New and emerging technologies are presenting the construction industry with a broad range of transformative opportunities for enhanced business growth, operational efficiencies, and much more. However, with these new opportunities come new challenges and risks. Challenges in the implementation of the emerging technologies include the following [11]:

- *Resistance to Change:* The construction industry is fraught with resistance to change. This has resulted in productivity growth that has lagged other sectors. Stakeholders in the construction industry have been reluctant to follow the digital world in adopting innovations. The attitude obscures their understanding of the technology. As a result, continuous demand for hardware and software upgrades has been a significant challenge for implementing several emerging technologies. However, modern-day professionals are tech-savvier, and more open to the new technologies the market has to offer.

- *Privacy:* As the digital landscape evolves, construction companies must navigate a complex web of regulations and consumer expectations. Imagine a world where your clients can trust that their personal information is handled with the utmost care, from the initial design consultation to the final building inspection.

- *Collaboration:* The construction technology wave is finally here and collaboration tools are just the tip of the iceberg. Construction collaboration and remote monitoring tools moved to the forefront as the need for such tools had never been more obvious. In today's fast-paced world, simply adding more people to a project team is not always enough to ensure efficient collaboration.

- *Sustainability:* As the world grapples with the pressing issues of climate change and environmental stewardship, the construction industry has a unique opportunity to lead the charge. Imagine a construction site where the materials we use are not just strong and durable, but also eco-friendly and recyclable. Picture a construction site where the energy we consume is not just clean, but also cost-effective, ensuring that sustainability and profitability can coexist in perfect harmony. Although the path to a sustainable future may seem daunting, but the construction industry is embracing the challenge with open arms.

- *Labor Shortages:* These have increased demand for labor and time-saving technologies, and recent advances have given rise to a new generation of more efficient, flexible, and adaptive construction technologies.

- *Costs:* It is widely noticed that the adoption of these innovations can minimize costs related to labor and injuries. At the same time, these technologies are considered expensive and the cost includes the technology systems, software, skilled engineers, and training. Successfully adopting the new technologies will require that companies evaluate their costs, benefits and risks and update construction practices as needed. For example, 3D printers are expensive, and the purchase price does not include materials or maintenance. Quality in 3D printing could become a costly mess if not constantly monitored. Companies who use the new technology will have to repeatedly pay the high price of a software system applying for security patches to keep their software efficient.

- *Interoperability:* When it comes to interoperability between BIM systems and augmented reality (AR) and virtual reality (VR) models, it is difficult to automatically update BIM models and construction schedules from AR and VR systems. Thus, the lack of integration between data standards makes it challenging to integrate BIM data photogrammetry and VR platforms.

- *Cloud Access:* Cloud access is basically over the Internet with technologies like cloud computing and the Internet of Things (IoT). Hence, to maximise the benefits of cloud solutions on the construction site, Internet connectivity must be available all the time. Poor network connectivity is among the challenges to cloud adoption by the construction industry.

1.6 CONCLUSION

The belief that the construction industry lags in adopting modern technology no longer holds true. Organizations like the American Institute of Constructors (AIC) recognize that to succeed in today's fast-paced environment, it is important to understand how technology has transformed the responsibilities of construction managers. The emerging technologies are bringing greater efficiency, productivity, and safety to the construction process. By leveraging emerging technologies, construction companies can save time and money, reduce waste, and improve overall project results.

The construction industry is experiencing a wave of emerging technologies that promise to transform the way things are built. Although some of these emerging technologies are still in their infancy, they offer immense potential for the future of construction. In the future of construction, technology reigns supreme and innovation knows no bounds. More information about emerging technologies in the construction industry can be found in the books [12-14].

REFERENCE

[1] "The seven coolest construction technologies for 2023," January 2023,

https://www.viewpoint.com/blog/seven-coolest-construction-technology-2023

[2] "5 Emerging technologies in the construction industry,"

https://www.worldconstructiontoday.com/news/5-emerging-technologies-in-the-construction-industry/

[3] M. N. O. Sadiku, P. A. Adekunte, and J. O. Sadiku, "Emerging technologies in the construction industry," *International Journal of Trend in Scientific Research and Development*, vol. 8, no. 6, November-December 2024, pp. 1142-1150.

[4] M. Halaweh, "Emerging technology: What is it?" *Journal of Technology Management & Innovation*, vol. 8, no. 3, 2013, pp. 108-115.

[5] N. Duggal, "Top 18 new technology trends for 2023," July 2023,

https://www.simplilearn.com/top-technology-trends-and-jobs-article

[6] C. Soto, "Construction technology: 16 building technologies driving the future," May 2024,

https://openasset.com/blog/construction-technology/

[7] S. Ahmed, "3 Game-changing construction technology trends to keep an eye on," July 2021,

https://tinyurl.com/yd6s7c93

[8] "Modern construction technologies transforming the industry," August 2023,

https://www.planradar.com/sa-en/modern-construction-technologies/

[9] J. Goyal, "10 Emerging technologies transforming the construction industry in 2024," March 18, 2023

https://www.novatr.com/blog/emerging-construction-technology-trends

[10] S. Coronado, "How emerging construction industry technology is shaping the future," July 2024,

https://www.ebacon.com/business-tips/how-emerging-construction-industry-technology-is-shaping-the-future/#:~:text=Real%2DTime%20Data%20and%20Analytics,driven%20decisions%20about%20staffing%20needs.

[11] E. Kissi, C. Aigbavboa, and E. Kuoribo, "Emerging technologies in the construction industry: Challenges and strategies in Ghana," January 2022,

https://www.researchgate.net/profile/Ewald-Kuoribo/publication/358357275_Emerging_technologies_in_the_construction_industry_challenges_and_strategies_in_Ghana_Challenges_and_strategies_in_Ghana/links/6429465e92cfd54f844a0596/Emerging-technologies-in-the-construction-industry-challenges-and-strategies-in-Ghana-Challenges-and-strategies-in-Ghana.pdf

[12] J. B. de Lartigue, *How to Surf the Waves of Change in the Construction Industry?: Advanced Construction Management: BIM Implementation Methodology, Technologies, and Leadership to Deliver Sustainable Buildings*. Kindle Edition, 2024.

[13] J. J. Hannon, *Emerging Technologies for Construction Delivery*. Transportation Research Board, 2007.

[14] U. Vanka, *The Future Is BIG: How Emerging Technologies are Transforming Industry and Societies*. Business Expert Press, 2023.

CHAPTER 2
ARTIFICIAL INTELLIGENCE IN CONSTRUCTION

"Artificial intelligence is poised to transform the construction industry, enabling machines to perform tasks better and faster than humans, leading to improved efficiency, cost reduction, and enhanced project outcomes."

—Anonymous

2.1 INTRODUCTION

The term artificial intelligence (AI) is used since the 1950s of the last century. It is an aggregative term for describing when a machine mimics human cognitive functions, like problem-solving, pattern recognition, and learning. It can be regarded as the ability of a digital computer or computer-controlled robot to perform tasks commonly associated with intelligent beings. As a branch of computer science, AI drives computers to sense and learn inputs like human-being for perception, knowledge representation, reasoning, problem-solving, and planning. You may have experienced AI in a voice search with Amazon's Alexa or Apple's Siri. Figure 2.1 shows AI symbol [1]. Artificial intelligence (AI) has become unavoidable for most industries, with construction being no exception.

Figure 2.1 AI symbol [1].

Construction is one of the least digitalized industries in the world. The construction industry is undergoing a significant transformation, driven by the rapid adoption of digital technologies. At the heart of this transformation is AI. The construction industry in North America is increasingly adopting AI-driven solutions to improve operational efficiency, project management, and safety. The key factors driving growth include the demand for sustainable and smart construction practices, the need for cost reduction, and a supportive regulatory environment. AI-powered solutions encompass a range of technologies and tools designed specifically for the construction industry. These solutions typically include machine learning algorithms, computer vision, natural language processing, robotics, and data analytics platforms. These solutions have successfully been applied in other industries to achieve increased profitability, efficiency, safety and security.

Construction activities contribute a lot to our society economically. The construction industry is undergoing a significant transformation, driven by the rapid adoption of digital technologies. Among these, artificial intelligence (AI) stands out as a powerful tool reshaping how construction projects are planned, executed, and managed. Artificial intelligence is where machines exhibit their own intelligence through using algorithms to solve problems.

Construction is growing at a rapid pace as digitalization continues and AI stands to deliver a number of advantages to the industry.

Artificial intelligence has become integral to construction processes, offering efficiency, precision, and innovation. From optimizing the design process to reducing energy consumption, AI has significant implications for the construction industry, making it more efficient and sustainable. AI in construction helps the industry as a whole overcome some of our toughest challenges, including safety concerns, labor shortages, cost, and schedule overruns [2].

This chapter aims to unravel AI applications in modern construction industry. It begins with describing what artificial intelligence is all about. It explains generative AI. It explains AI in construction and some of its applications. It highlights the benefits and challenges of AI in the construction industry. It concludes with comments.

2.2 WHAT IS ARTIFICIAL INTELLEGENCE?

The term "artificial intelligence" (AI) is an umbrella term John McCarthy, a computer scientist, coined in 1955 and de-fined as "the science and engineering of in-telligent machines." It refers to the ability of a computer system to perform human tasks (such as thinking and learning) that usually can only be accomplished using human intelligence [3]. Typically, AI systems demonstrate at least some of the following human behaviors: planning, learning, reasoning, problem solving, knowledge representation, perception, speech recognition, decision-making, language translation, motion, manipulation, intelligence, and creativity.

The 10 U.S. Code § 2358 define artificial intelligence as [4]:

1. "Any artificial system that performs tasks under varying and unpredictable circumstances without significant human oversight, or that can learn from experience and improve performance when exposed to data sets.

2. An artificial system developed in computer software, physical hardware, or other context that solves tasks requiring human-like perception, cognition, planning, learning,

communication, or physical action.

3. An artificial system designed to think or act like a human, including cognitive architectures and neural networks.

4. A set of techniques, including machine learning, that is designed to approximate a cognitive task.

5. An artificial system designed to act rationally, including an intelligent software agent or embodied robot that achieves goals using perception, planning, reasoning, learning, communicating, decision making, and acting."

AI provides tools creating intelligent machines which can behave like humans, think like humans, and make decisions like humans. The main goals of artificial intelligence are [5]:

1. Replicate human intelligence
2. Solve knowledge-intensive tasks
3. Make an intelligent connection of perception and action
4. Build a machine which can perform tasks that requires human intelligence
5. Create some system which can exhibit intelligent behavior, learn new things by itself, demonstrate, explain, and can advise to its user.

AI is not a single technology but a range of computational models and algorithms. The concept of AI is an umbrella term that encompasses many different technologies. AI is not a single technology but a collection of techniques that enables computer systems to perform tasks that would otherwise require human intelligence. The major disciplines in AI include [6]:

- *Expert systems*
- *Fuzzy logic*
- *Neural networks*
- *Machine learning (ML)*

- *Deep learning*
- *Natural Language Processors (NLP)*
- *Robots*

These computer-based tools or technologies have been used to achieve AI's goals. Each AI tool has its own advantages. Using a combination of these models, rather than a single model, is recommended. Figure 2.2 shows a typical expert system, while Figure 2.3 illustrates the AI tools. These tools are gaining momentum across every industry. Analytics can be considered a core AI capability.

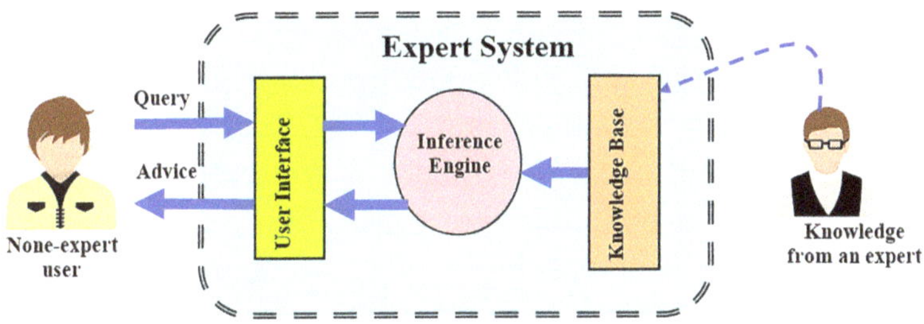

Figure 2.2 A typical expert system.

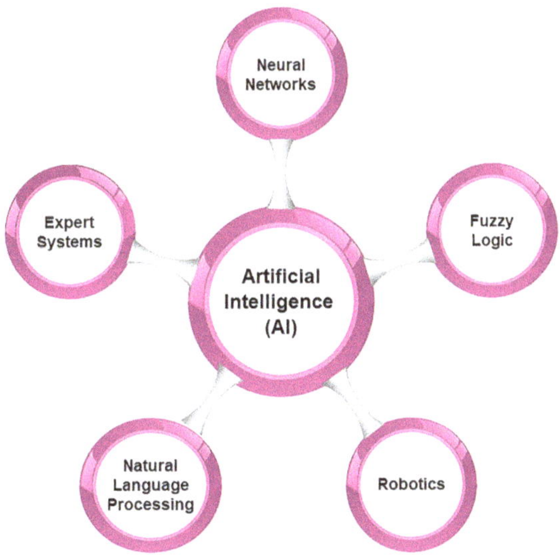

Figure 2.3 AI tools.

2.3 GENERATIVE AI

Artificial Intelligence (AI) is increasingly a part of our world and it is rapidly changing our lives. Generative AI (GenAI) is a subset of artificial intelligence that uses generative models to produce text, images, videos, or other forms of data. Generative AI (GenAI) is a term for any type of AI system capable of using generative models to create new forms of humanlike creative content, like text, images, music, audio, video and more. GenAI models include various algorithms able to learn the various patterns and structures of input training data before generating novel outputs with similar characteristics. It is essentially a narrow type and application of the broader artificial intelligence umbrella of technologies. It describes algorithms (such as ChatGPT) that can be used to create new content, including audio, code, images, text, simulations, and videos. It is specifically designed and trained to generate new content. The versatility and potential of GenAI to transform various aspects of business operations make it an attractive investment for companies across industries. GenAI uses neural networks, machine learning, deep learning models, complex algorithms, and large and varied training datasets to produce original content based on user input and how to reason in ways akin to a human brain. The technology is built on AI tools shown in Figure 2.4 [6]. It uses neural networks to identify the patterns and structures within existing data to generate new and original content.

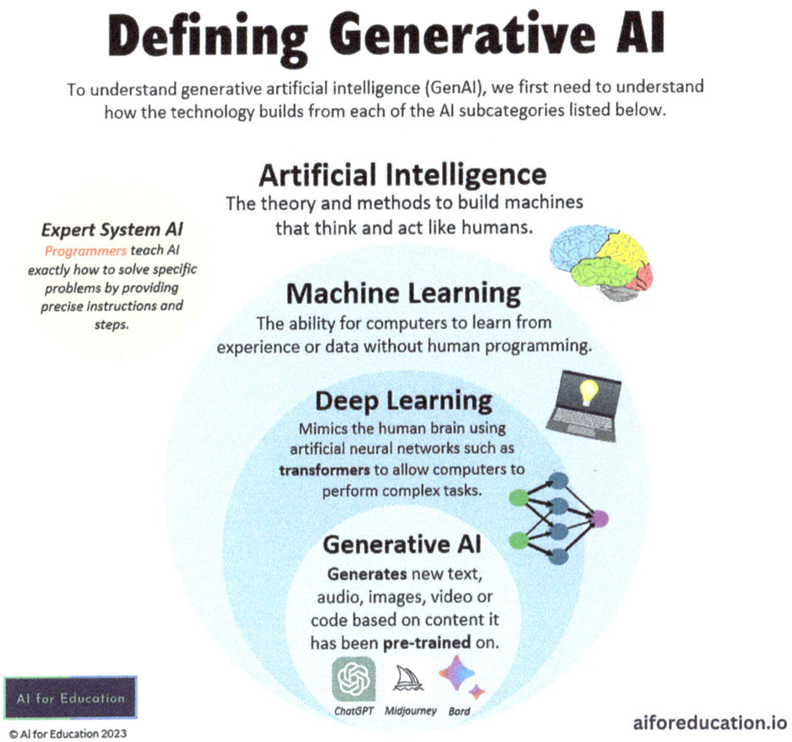

Figure 2.4 GenAI built on AI tools listed above [5].

Generative AI can be thought of as a machine-learning model that is trained to create new data, rather than making a prediction about a specific dataset. Since its inception, the field of machine learning used both discriminative models and generative models, to model and predict data. A generative AI system is constructed by applying unsupervised machine learning or self-supervised machine learning to a data set. The most common way to train a generative AI model is to use supervised learning. Generative AI can also be trained on the motions of a robotic system to generate new trajectories for motion planning or navigation. Generative AI models are used to power chatbot products such as ChatGPT [8].

Generative AI is transforming nearly all aspects of the pharmaceutical industry, revamping the way companies operate and potentially unlocking billions of dollars in value. The pharmaceutical-operations value chain encompasses sourcing, manufacturing, quality, and the supply chain—and gen AI is expected to improve them all

2.4 AI IN CONSTRUCTION

At its core, construction is an inherently human enterprise, a domain where the personal touch, expertise, and innovative spirit of individuals are irreplaceable. Construction, as a large sector of the economy, plays prominent roles in driving economic growth and long-term national development. Globally, individuals and businesses spend over $10 trillion per year on construction-related activities. Construction projects create a broad range of job opportunities for 7% of the world's working population. Despite its economic importance, an obvious issue is poor labor productivity during the construction procedure, negatively leading to the waste of manpower, material resources, and financial resources. Figure 2.5 shows construction site [9], while Figure 2.6 shows construction workers [10].

Figure 2.5 A typical construction site [9].

Figure 2.6 Construction workers [10].

The growth of the construction industry is severely limited by the myriad complex challenges it faces such as cost and time overruns, health and safety, productivity and labor shortages. Also, construction industry is one the least digitized industries in the world, which has made it difficult for it to tackle the problems it currently faces [11].

AI in construction is an umbrella term that describes a wide range of AI-powered tools and machinery designed specifically for use in the construction industry. It includes specialized software that utilizes machine learning algorithms as well as advanced industrial robots. AI in the construction industry has emerged as a technology that is poised to transform the industry. Its components are depicted in Figure 2.7 [12].

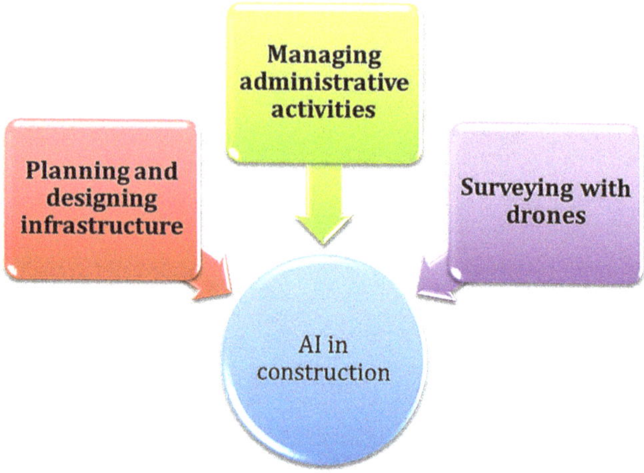

Figure 2.7 Components of AI in construction [12].

It has the potential to help players realize value throughout project lifecycles, including design, bidding, and financing; procurement and construction; operations and asset management; and, business model transformation. AI can help construction companies to optimize the use of resources, manage timelines, and overall achieve better efficiency. Imagine an industry where project delays are predicted with uncanny accuracy and workplace injuries are anomalies. Construction AI companies are key drivers of this revolution. Figure 2.8 shows a typical use of AI in construction [13].

Figure 2.8 A typical use of AI in construction [13].

To make AI live up to expectations, more and more companies are actively investing in various AI technologies. Major companies operating in the AI in construction market include [14]:

- Alice Technologies Inc.
- Autodesk, Inc.
- Bentley Systems Inc.
- Dassault Systemes SE
- IBM Corporation
- Microsoft Corporation
- NVIDIA Corporation
- Oracle Corporation
- Predii
- PTC Inc.
- SAP SE
- Smartvid.io Inc.

2.5 APPLICATIONS OF AI IN CONSTRUCTION

Artificial Intelligence is concerned with smart machines or a system due to which machines are capable of solving engineering problems and performing the various tasks that require human intelligence. It is currently revolutionizing industries such as manufacturing, retail, and telecommunications. The construction industry has been adopting AI technologies to automate and optimize processes, improve site safety and security, and extract insights from complex data. The potential applications of AI in construction are vast. Common application areas of AI in construction include the following [9,15]:

1. *Construction Engineering and Management (CEM):* Construction projects are recently becoming larger and more complex, i.e. encompassing more and more participants focusing on diverging interests while the given frames of time and budget

are getting tighter. Construction management is used to establish an efficient organization of all these issues and able to predict the result with a high degree of precision and certainty.

The construction engineering and management (CEM) inside the scope of the architecture, engineering, and construction (AEC) industry is fraught with its own problems and complications, which covers a set of construction-related activities and processes along with human factors and interactions. With the extensive adoption of artificial intelligence (AI), construction engineering and management is experiencing a rapid digital transformation. CEM is going through constant innovations towards digitalization and intelligence, in order to realize a considerable boost in automation, productivity, and reliability. For the purpose of launching the real digital strategies in CEM, artificial intelligence (AI) acts as the backbone to change the way a construction project performs. There is immense interest in implementing a variety of AI methods in the CEM domain to seize the valuable opportunity of digital evolution for better performance and profitability.

2. *Project Management:* Thus is one primary area of construction where AI can be quite effective. Managing a large construction project can challenge even the most experienced teams. AI technologies are being leveraged to streamline and optimize various aspects of project management. The growth of AI in project management is revolutionizing how construction projects are planned, executed, and managed, leading to improved outcomes and reduced project risks. AI algorithms can analyze vast amounts of data from past projects to predict potential delays, identify risks, and recommend efficient scheduling. AI software designed for construction project management captures photo and video footage on work sites and automatically generates relevant insights, helping construction management track work progress and crew productivity in real-time. AI-powered tools are becoming integral to the design phase of construction projects. Generative design, an AI-driven process, allows architects and engineers to input specific design goals and constraints, such as materials, budget, and building codes, and receive a range of optimized

design options.

3. *Estimations and Bidding:* This involves distributing the plans to trades, contractors, and manufacturers for bidding. Construction estimations and bidding have historically been highly manual and time-consuming processes. Once architects and developers come up with a design, a general contractor uses the blueprint to calculate a bid. They then repeat the process with subcontractors: plumbers, electricians, framers, etc. Putting every individual bid together is a laborious process without guaranteed returns for the general contractor.

4. *Robotics:* AI robotics is becoming an increasingly important and rapidly growing field of construction technology. Robotics in the construction sector is taking off. These robots are typically autonomous machines capable of automating tasks on construction sites. AI robots that are already present on sites include: 3D printers, bricklaying robots, painting robots, welding robots, layout drawing robots, rebar tying robots, monitoring drones and rovers, self-operating heavy equipment, and self-driving trucks. Now you can retrofit various heavy equipment machines with artificial intelligence to perform duties like a human worker. AI robots complement the work of human workers, and make it safer and more efficient. For example, bricklaying robot works alongside masons and only completes the literal heavy-lifting part of the bricklaying process. One will find industrial robots in warehouses and construction sites where repetitive tasks are required. The robot "Spot" has proven to be truly disruptive for the global construction industry, with a number of companies using it on their construction sites. Figure 2.9 shows some robots working along with human workers [16].

Figure 2.9 Robots working along with human workers [16].

5. *Design and Planning:* AI technologies like machine learning and generative design algorithms can streamline decision-making processes for architects, engineers, and construction professionals involved in design and planning. AI can analyze complex data, including building codes, zoning laws, topographical information, and historical projects, to generate designs that are in line with regulatory requirements.

6. *Predictive Maintenance:* This involves using AI algorithms to monitor equipment health and forecast failures. It helps construction companies extend the lifecycle of machinery, improve on-site safety, and prevent project downtimes. AI can be used to analyze data from building systems, such as HVAC and electrical systems, to predict when maintenance will be needed and prevent equipment failure. Construction equipment and machinery are critical assets that require regular maintenance to avoid breakdowns and costly downtime. AI tools can analyze data from sensors installed on machinery to predict when maintenance is needed, based on factors such as usage patterns, wear and tear, and environmental conditions. This shift from reactive to predictive maintenance can help construction firms minimize downtime, reduce repair costs, and extend equipment lifespan. This is how

predictive maintenance works in construction. Machine learning algorithms monitor equipment health by analyzing sensor data (including information about environmental conditions and equipment performance stats) and detect anomalies in real-time. They then diagnose issues and alert operators before critical failures occur, suggesting appropriate preventative maintenance action. For example, suppose a construction company is building a large infrastructure project, like a bridge. With the help of machine learning (ML), they can use predictive maintenance to keep the construction equipment in top shape, reducing the risk of unexpected breakdowns and delays. The company can use ML for quality control by analyzing images and videos of the construction site to ensure that all the materials and workmanship meet the required standards.

7. *Automating Tasks:* AI has been found to automate and accelerate the process of learning, reasoning, and perceiving from large data. AI drives the process of project management more technically automatable and objective. The use of AI-powered drones and robotics in construction is gaining traction. These trends are driving the industry toward smart and automated construction practices, leading to increased efficiency, productivity, and quality in construction projects. The latest tools help automate mundane tasks like detecting title blocks and sheet names, freeing professionals to focus on more strategic aspects of their work. Technologies aided by computer vision mainly contribute to the automatic and robust vision- based inspection, which gradually take the place of the error-prone, time-consuming, laborious, and dangerous manual observation by people. Computer vision supports the automated process of detecting and assessing the defects and damage (i.e., crack, spalling, corrosion, holes, joint damage, etc.) exiting on various types of civil infrastructure, including buildings, bridges, tunnels, roads, sewer pipes, etc. AI-powered platforms automate routine administrative tasks, such as scheduling meetings or generating reports, freeing project managers to focus on more strategic aspects of their roles. AI-powered construction vehicles such as bulldozers, excavators, and dump trucks can work independently, freeing laborers to focus on

other tasks.

8. *Risk Mitigation:* Every construction project has some risk that comes in many forms such as quality, safety, time, and cost risk. AI can monitor, recognize, evaluate, and predict potential risk in terms of safety, quality, efficiency, and cost across teams and work areas even under high uncertainty. Various AI methods can effectively address the limitations of traditional risk analysis, such as the vagueness and vulnerability from specialist experience and subjective judgment. As a result, the AI-based risk analysis can provide assistive and predictive insights on critical issues. AI presents valuable opportunities to realize early troubleshooting to prevent undesirable failure and accidents in the complex workflow. Robots can take charge of unsafe activities to minimize the number of humans working in dangerous environments. AI-powered robots have been directly adopted on the construction site to take over the repetitive and routine construction tasks, such as bricklaying, welding, tiling, and others. They can work continuously without taking a break at almost the same rate and quality.

2.6 BENEFITS

AI enhances safety in construction operations through real-time monitoring, hazard detection, and compliance enforcement. It also enables quality control and defect detection, ensuring high construction standards and reducing rework. The growth of AI solutions in construction is driven by their ability to automate processes, optimize resource allocation, enhance safety, and improve decision-making. It can sometimes be overwhelming to manage an entire project single handed, but with the assistance of AI, it is becoming easier. Other benefits include [15,17]:

- *High Efficiency:* One of the most significant advantages of AI is increased efficiency. AI tools can process data much faster than humans, enabling real-time decision-making that can reduce delays and cut costs. Another important use of AI techniques is in optimization problems, aiming to make the construction project run more smoothly and efficiently. Tactical decisions can be made for trouble-shooting at an early stage, driving the improvement of operational efficiency.

- *Predictive Analytics:* AI algorithms can analyze historical data and project future outcomes, enabling construction companies to optimize scheduling, resource allocation, and budget planning.

- *Prevent Cost Overruns:* Most mega projects go over budget despite employing the best project teams. Artificial neural networks are used on projects to predict cost overruns based on factors such as project size, contract type, and the competence level of project managers. AI helps staff remotely access real-life training material which helps them enhance their skills and knowledge quickly. As a result, project delivery is expedited.

- *Project Planning:* One construction intelligence company launched in 2017 with the promise that its robots and artificial intelligence hold the key to solving late and over budget construction projects. Algorithms of the future will use an AI technique known as "reinforcement learning." This technique allows algorithms to learn based on trial and error. It can assess endless combinations and alternatives based on similar projects. It aids in project planning since it optimizes the best path and corrects itself over time.

- *Safety:* Safety is a top priority in the construction industry. AI plays a significant role in enhancing on-site safety. Construction workers are killed on the job five times more often than other laborers. According to OSHA, the leading causes of private sector deaths (excluding highway collisions) in the construction industry were falls, followed by struck by an object, electrocution, and caught-in/between. Computer vision technologies, a subset of AI, monitor construction sites in real-time, identifying safety hazards such as workers not wearing protective gear or machinery operating outside safe parameters.

- *Labor Shortage:* This happens more often in the construction industry. It could be due to various factors such as an aging workforce, declining interest in the industry, or a growing economy leading to more job opportunities in other fields. Labor shortage and a desire to boost the industry's low productivity are compelling construction firms to invest in AI and data science. Construction companies are starting to use AI and machine learning to better plan for distribution of labor and machinery across jobs. A robot constantly evaluating job progress and the location of workers and equipment enables project managers to tell instantly which job sites have enough workers and equipment to complete the project on schedule. Despite the predictions of massive job losses, AI is unlikely to replace the human workforce. Instead, it will alter business models in the construction industry, reduce expensive errors, reduce worksite injuries, and make building operations more efficient.

- *Post-construction:* Building managers can use AI long after construction is complete. By collecting information about a structure through sensors, drones, and other wireless technologies, advanced analytics and AI-powered algorithms gain valuable insights about the operation and performance of a building, bridge, roads, and almost anything in the built environment.

- *Sustainability:* AI can improve the overall sustainability of the construction industry. AI tools can optimize the use of resources and reduce waste, helping construction companies adopt more sustainable construction practices. As concerns about climate change and resource scarcity grow, the construction industry faces increasing pressure to adopt sustainable practices. AI can assist in this effort by optimizing material use, reducing waste, and improving energy efficiency.

- *Better Project Planning:* AI systems can streamline scheduling and resource allocation, optimizing the profitability and timeliness of large projects. AI can also recognize structural and functional issues in models, allowing for early course correction and reducing the risk of reworks and waste.

- *Quality Control:* AI-powered cameras and sensors can monitor the quality of construction work, identify defects, and alert workers and managers to issues that need to be addressed.

- *Enhanced Creativity:* AI-powered software assists architects and engineers to think beyond the ordinary. By analyzing vast data sets, AI-driven software generates alternative design options, like digital brainstorming.

• *Recruitment Process:* The use of AI in the recruitment process has shown significant improvements and results. By removing human bias, organizations can make fairer and more objective hiring decisions, leading to a more diverse and inclusive workforce. Using AI in the recruitment process improves efficiency and effectiveness and promotes fairness and diversity within organizations. It helps organizations achieve fair, unbiased, and inclusive hiring by removing human bias, expanding the talent pool, enhancing innovation, conducting objective assessments, automating interviews, and providing personalized onboarding.

• *Project Delays:* The construction industry often faces numerous risks and uncertainties that can impact project timelines and budgets. By adopting predictive analytics with AI technology, organizations can better identify and mitigate potential project delays.

• *Productivity:* With data like job records, cost estimates, project plans, camera footage, and drone surveys all accessible via the cloud, it is possible to analyze this data and take actions that boost productivity. AI will help by ingesting these disparate datasets, providing recommendations that reduce cost and increase efficiency.

• *Soil Moisture:* With the help of AI and machine learning in construction, the moisture present in the soil can be estimated with the help of remotely sensor data or even images taken from the satellite with increased accuracy.

2.7 CHALLENGES

While acknowledging the benefits of AI applications, numerous challenges which are relevant to AI still exist in the construction industry. Construction with inherent complexity is regarded as one of the most dangerous industries, which is greatly susceptible to a variety of unpredictable factors, such as participants in different roles, the changeable environment in large uncertainty, struck-by-equipment hazard, etc. Other challenges include [15,18]:

• *Resistance to Change:* One primary obstacle of AI is the industry's inherent resistance to change. Construction has traditionally been slow to adopt new technologies, often due to concerns about cost, complexity, and potential disruption to established workflows. Many construction companies, particularly smaller ones, may lack the resources or expertise to implement AI effectively. In order to successfully adopt AI technology in the construction industry, it is crucial to cultivate a workforce that is ready for change. One of the first steps in achieving this is to develop tailored training programs for employees. In addition to training programs, implementing employee engagement programs can greatly contribute to cultivating a workforce ready for change. These programs foster employees' sense of belonging, motivation, and loyalty, ultimately driving their willingness to embrace new technologies such as AI.

- *High Cost:* The high cost of investment and operation is a significant obstacle hampering AI in construction market growth. Developing and implementing AI technologies in construction projects require substantial financial resources including the cost of acquiring AI systems, hardware, software, and specialized expertise. The high cost of investment and operation makes it challenging for small- and medium-sized construction companies to adopt AI solutions. While the long-term savings and efficiency gains can outweigh these costs, the initial financial outlay can be a deterrent. Figure 2.10 shows the cost of implementing AI in construction [1].

Figure 2.10 Cost of implementing AI in construction [1].

- *Ethical Concern:* AI-based solutions can pose ethical and legal risks, such as biases, discrimination, accountability, and transparency. Governments play a crucial role in ensuring responsible and reliable adoption of this technology across all sectors.

- *Privacy and Security:* Artificial intelligence thrives on robust datasets. To operate effectively, companies must ensure data quality, accuracy, and accessibility. Sharing data can incur risks, including cyberattacks and privacy breaches.

- *Fatal Accidents:* The construction industry tends to cause a small scale of fatal accidents with higher frequency than other domains, which is even responsible for 30-40% of fatalities worldwide, emphasizing the necessity of construction management for safety guarantee and accident prevention. Construction in China has been regarded as one of the riskiest industries, where the number of fatal accidents exceeds many developed countries without a significant downward trend.

- *Uncertainty:* In the construction phase, a great deal of unknown uncertainty comes from the ground conditions, soil-structure interaction, weather conditions, building material properties, design changes, reliability of suppliers, and others. Uncertainties are unknown before they occur, which can be regarded as unavoidable threats to raise the risk of project failure. Notably, a high level of uncertainty is inherent in complicated construction projects, which is closely related to various factors. The improper estimation of uncertainty will impede the progress of the project.

- *Workforce Adaptation:* This is another significant challenge. The integration of AI tools requires a workforce skilled in construction and comfortable using advanced digital technologies. Ensuring that employees are trained to use AI tools is crucial to maximizing their benefits. This includes technical training and a shift in mindset, where workers see AI as a tool to enhance their capabilities rather than a threat to their jobs.

- *Lack of Infrastructure:* This is another area of concern. Many construction sites, particularly those in remote locations, may lack the necessary connectivity and hardware to support AI solutions effectively. Construction companies need to invest in upgrading their infrastructure to fully leverage AI capabilities.

- *Aging Workforce:* The aging workforce is one more challenge that can hinder the implementation of AI in construction. The construction industry faces serious problems to recruit young workers. The good news is that the rise of digital technologies can help the sector redefining itself and turn to a promising destination for an ambitious and tech-savvy workforce. The UK government has started to focus its efforts on bringing more workers in the industry by 2025, while also investing in training programs for the industry's workforce. Figure 2.11 shows construction workforce by age in the United Kingdom [19].

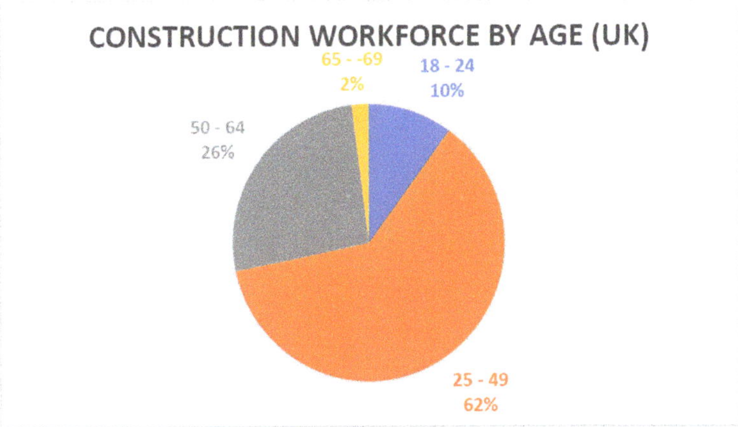

Figure 2.11 Construction workforce by age in the United Kingdom [19].

2.8 CONCLUSION

The construction industry, generally regarded as a laggard in adoption of cutting-edge technology, has surprisingly been quick to warm up to AI. Recent research indicates that applications of AI technologies are increasingly permeating the construction industry in all areas of architectural design, engineering design, and construction services. As AI continues to evolve, its influence on the construction sector is becoming more profound, bringing about changes that enhance efficiency, safety, and innovation. When AI talents continue to mature, it is believed that AI methods will become the next digital frontier to easily transform massive data into useful knowledge, leading to a high degree of automation and intelligence in both industry and commerce.

One can expect AI to play a more significant role in construction in the coming years. The future of artificial intelligence in the construction industry looks promising, with new developments continuously emerging. In the immediate future, the construction industry is projected to increase more focus and investment in AI. The construction industry may find itself increasingly reliant on AI solutions to stay competitive in a rapidly changing landscape. For this reason, forward-looking contractors should keep an eye on AI. More information on artificial intelligence in the construction industry is available from the books in [20-23] and the following related journals:

- *Energy and AI*
- *The AI Journal*
- *AI Magazine*
- *Journal of Intelligence*
- *Journal of Building Engineering*
- *Journal of Construction Engineering and Management*
- *Automation in Construction*

REFERENCE

[1] "The rise of artificial intelligence in construction industry," June 2022,

https://gharpedia.com/blog/artificial-intelligence-in-construction/

[2] M. N. O. Sadiku, P. O. Adebo, and J. O. Sadiku, "Artificial intelligence in the construction industry," *International Journal of Trend in Research and Development*, vol. 12, no. 1, January-February 2025, pp. 40-47.

[3] M. N. O. Sadiku, "Artificial intelligence," IEEE Potentials, May 1989, pp. 35-39.

[4] "Artificial intelligence (AI),"

https://www.law.cornell.edu/wex/artificial_intelligence_(ai)

[5] "Artificial intelligence tutorial,"

https://www.javatpoint.com/artificial-intelligence-tutorial

[6] D. Quinby, "Artificial intelligence and the future of travel," May 2017,

https://www.phocuswright.com/Travel-Research/Research-Updates/2017/Artificial-Intelligence-and-the-Future-of-Travel

[7] "Generative AI explainer," Unknown Source.

[8] M. N. O. Sadiku, P. A. Adekunte, and J. O. Sadiku, "Generative artificial intelligence," *International Journal of Trend in Scientific Research and Development*, vol. 8, no. 6, November-December 2024, pp. 561-570.

[9] "AI in construction in 2024 and beyond: Use cases and benefits,"

https://www.tribe.ai/applied-ai/ai-in-construction

[10] "Generative AI in the construction sector: Taking building technology to new heights," May 2023,

https://www.forbes.com/sites/forbesbooksauthors/2023/05/02/

generative-ai-in-the-construction-sector-taking-building-technology-to-new-heights/

[11] O. Oyedele et al., "Artificial intelligence in the construction industry: A review of present status, opportunities and future challenges," *Journal of Building Engineering*, vol. 44, December 2021.

[12] N. Joshi, "These 3 applications of AI will reshape the construction industry," September 2018,

https://www.linkedin.com/pulse/3-applications-ai-reshape-construction-industry-naveen-joshi/

[13] B. Cooper, "Best construction AI companies: Top 10 transforming the industry," August 2024,

https://downtobid.com/blog/construction-ai-companies

[14] "Private AI infrastructure solves for privacy and regulatory compliance requirements,"

https://engage.equinix.com/leaders-challenge/items/private-ai-infrastructure-solves-for-privacy-and-regulatory-compliance-requirements-11b3

[15] Y. Pan and L. Zhang, "Roles of artificial intelligence in construction engineering and management: A critical review and future trends," *Automation and Construction*, vol. 122, February 2021.

[16] "Are AI and robotics starting to revolutionize the construction industry?" July 2024,

https://www.geoweeknews.com/blogs/artificial-intelligence-construction-ai-technology-robotics-bim

[17] S. Rao, "The benefits of AI in construction," April 6, 2022,

https://constructible.trimble.com/construction-industry/the-benefits-of-ai-in-construction

[18] "The rise of artificial intelligence in construction,"

https://construction-today.com/news/the-rise-of-artificial-intelligence-in-construction/#:~:text=AI%2Ddriven%20robots%20and%20drones,error%20and%20increasing%20overall%20productivity.

[19] "Artificial intelligence: The future of construction,"

https://www.letsbuild.com/blog/artificial-intelligence-the-future-of-construction

[20] L. Zhang et al., *Artificial Intelligence in Construction Engineering and Management*. Springer, 2021.

[21] T. Dinesh, *Artificial Intelligence in Construction Industry. Kindle Edition*, 2020.

[22] E. Sudjono, *Artificial Intelligence in Construction Management: Evolutionary Fuzzy Hybrid Neural Network as Decision-Making Machine*. VDM Verlag, 2008.

[23] D. H. de la Iglesia (ed.), *New Trends in Disruptive Technologies, Tech Ethics, and Artificial Intelligence. The DITTET 2024 Collection*. Springer, 2024.

CHAPTER 3
ROBOTICS IN CONSTRUCTION

"The future of construction isn't about robots replacing humans, but about humans and robots working hand-in-hand, with robots handling repetitive and dangerous jobs, allowing humans to focus on creativity and problem-solving."

—Anonymous

3.1 INTRODUCTION

The construction industry is one of the most important industrial sectors in North America. It has huge economic importance. The construction industry is a labor-intensive industry and is one of the least automated industries of all. It is highly complex, with a construction project usually including hundreds of tasks and multiple phases. The construction industry, traditionally seen as slow to embrace technological change, is undergoing a remarkable transformation. This transformation is not about replacing human labor; it is about augmenting and enhancing human capabilities, opening up new possibilities in building design and construction methods. Figure 3.1 shows a typical construction site [1], while Figure 3.2 shows some construction workers [2].

Figure 3.1 A typical construction site [1].

Figure 3.2 Construction workers [2].

Traditionally, construction works of the large building were performed by human labor in an uncomfortable working condition for a long duration. Today, construction demands more precise, faster, and cost-effective methods to build increasingly elaborate designs. To be effective requires computer-assisted planning, engineering, and a flexible automation process. There is where robots come in.

Robotics is the branch of engineering that integrates computer science with mechanical and electrical engineering. It is now finding its way into the construction industry. Robots usually speed up production and eliminate the cost of labor. Imagine a team of robots working along with laborers on a construction site to transform the landscape.

The construction industry is one of the largest and most labor-intensive industries in the world. It is currently grappling with issues related to quality, timelines, safety, and complex environments. The industry is on the verge of a digital revolution, using technology to tackle issues like productivity, safety, labor shortages, and collaboration. It is ready for more changes ahead, emphasizing digital transformation and the integration of new technologies to meet the evolving needs of society. The integration of robotics technology in construction facilitates construction professionals with quality-assured outcomes and reduced human errors. The modern robotic technology has the potential to revolutionize and provide numerous advantages to the construction professionals [3].

In this chapter, we will explore how robotics is making sweeping changes in the construction industry. It begins with explaining what robots are. It describes construction robots and presents some of their applications. It highlights the benefits and challenges of robotics in the construction industry. It concludes with comments.

3.2 WHAT ARE ROBOTS?

The word "robot" was coined by Czechriter Karel Čapek in his play in 1920. Isaac Asimov coined the term "robotics" in 1942 and came up with three rules to guide the behavior of robots and later added the zeroth law [4]:

- Law 0: A robot may not injure humanity or through inaction, allow humanity to come to harm.
- Law 1: Robots must never harm human beings,
- Law 2: Robots must follow instructions from humans without violating rule 1,
- Law3: Robots must protect themselves without violating the other rules.

Robots are becoming increasingly prevalent in almost every industry, from healthcare to manufacturing. Figure 3.3 indicates that robotics is one of the branches of artificial intelligence.

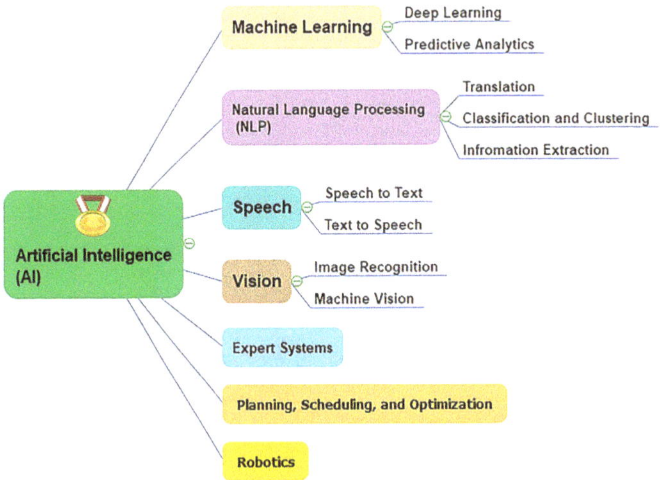

Figure 3.3 Robotics is one of the branches of artificial intelligence.

Although there are many types of robots designed for different environments and for different purposes/applications, they all share four basic similarities [5]:

1) All robots have some form of mechanical construction designed to achieve a particular task;
2) They have electrical components which power and control the machinery;
3) All robots must be able to sense its surroundings; a robot may have light sensors (eyes), touch and pressure sensors (hands), chemical sensors (nose), hearing and sonar sensors (ears), etc.
4) All robots contain some level of computer programming code.

Programs are the core essence of a robot since they provide intelligence. There are three different types of robotic programs: remote control, artificial intelligence, and hybrid. Some robots are programmed to faithfully carry out specific actions over and over again (repetitive actions) without variation and with a high degree of accuracy.

Robotics is an interdisciplinary field that involves the design, construction, operation, and use of robots. It is a branch of engineering and computer sciences that includes the design and use of machines that are capable of performing programmed tasks without human involvement. The field develops machines that can efficiently carry out various tasks, can automate tasks, and do various jobs that a human might not be able to do. Robots could someday be our drivers, companions, collaborators, teachers, specialists, and exploration pioneers [6].

3.3 CONSTRUCTION ROBOTS

With rising global populations and the consequent demand for infrastructure, the construction industry faces challenges that require solutions beyond traditional practices.

Robotics is poised to revolutionize construction methodologies, ensuring projects are not only completed faster but also with a level of precision hitherto unseen. Labor shortages, the quest for perfection, tight project timelines, and heightened safety standards have all accelerated the need for automation and robotics.

Construction robot refers to robotic devices that are developed for specific construction-related activities. We are living in the booming era of construction robots. Robots come in different types and sizes, but not all robots are suitable for construction. Construction robots can be roughly divided into three categories based on the controlling mechanism: programmable robots, teleoperated systems, and fully autonomous robots. They can also be classified based on several factors such as power source, shape, size, type of application, mechanical structure (mobile or fixed), workspace geometry, motion characteristics, and levels of autonomy. The following types of robots are commonly used in construction [7,8]:

1. *Demolition Robots:* Using robots for demolishing large structures and buildings can accelerate the demolition process, improve efficiency, and save both money and time. Robots make it easy to break down walls, collect debris, and crush the concrete in confined places. Using robots allows the operator to operate

by staying at a safe distance from the contaminants, debris, and crushing concrete, making it a safer option for human workers. They allow removing workers from some of the most dangerous activities due to the presence of noise, dust, and the risk of collapse or explosions. A demolition robot is shown in Figure 3.4 [9].

Figure 3.4 A demolition robot [9].

2. *Bricklaying Robots:* Bricklaying is the most monotonous and tedious task in the construction process. This increases the possibility of error. The bricklaying robots, however, enable us to execute such monotonous tasks efficiently. They can precisely place each brick, ensuring consistent quality and reducing the likelihood of human error.

Most of the bricklaying robots utilize arms of industrial robots to assemble the masonry structure of the building, while other human-operated robot mixes bonding agent or cement. The major advantage is efficiency and reduced construction time. Bricklaying robots can lay thousands of bricks in a day, ensuring that walls are straight and of consistent quality. These robots, guided by CAD designs, can adapt to changes in real-time, ensuring that any design modifications are immediately incorporated. The first bricklaying robot, named "Motor Mason," was patented in 1904. It required three people for operating. It was considered as five to 10 times faster than humans were.

3. *Welding Robots:* Robotic welding is critical to construction for high-quality welds in shorter cycle times. Robots can perform precise welding and maneuver easily in difficult locations. They are expensive to purchase, which means that an average business cannot afford one.

4. *Exoskeleton:* Exoskeletons, as the name suggests, are external skeletal systems that can be worn by humans. The exoskeleton is an extremely helpful device for a construction worker. It is basically a metallic framework that can be worn by a person. An exoskeleton assists a worker in lifting heavy loads, maintaining the body balance of that person. This mechanical suit can exponentially increase the strength, speed, and agility of an average worker. It allows any worker to lift and carry heavier objects than an average person.

The downside of this technology is the bulkiness, and most of them lack speed and agility. Due to the use of electric motors and heavy materials, exoskeletons are usually expensive, cumbersome, heavy, and uncomfortable to wear for a long period.

5. *Drones:* Drones are actually unmanned robots controlled remotely by operators to accomplish various complicated tasks. They can create real-time aerial images from the building objects and reveal assets and challenges presented by the terrain. They can provide convenient and smart ways of site supervision and management and result in better operations, planning, and effective on-site adjustments. They can improve safety and monitor project progress on construction sites. Flying mobile robots provide a significant advantage in construction by eliminating constraints associated with ground-based movement and the need for scaffolding. Figure 3.5 shows a typical drone [10].

Figure 3.5 A typical drone [10].

6. *Robotics Arms:* These are the backbone of manufacturing, machinery, and industrial automation. These are the most widespread robots in construction. Their main tasks include welding, handling materials, thermal spraying, painting, drilling, and waste separation.

7. *Mobile Robots:* Mobile robot construction systems provide cutting-edge solutions for accelerating construction. They can construct structures larger than their size, and their advantages on uneven terrain make them essential for numerous applications.

8. *Swarm Robots:* Swarm robots enable cooperation and simultaneous task execution, making them a cutting-edge solution. In construction, swarm robots are made up of many small robots to build a design without any guidance.

9. *Teleoperated Robots:* These robots assist humans in a variety of tasks from a distance. Unlike autonomous robots, teleoperated machines rely on some degree of human interaction to function. Teleoperation allows construction robots to be remotely controlled from a safe distance, whether from within the same city, across states, or even from other countries.

10. *Finishing Robots:* Among the different activities involved in building construction, finishing tasks such as plastering, panting, or coating application have seen little development in the field of

on-site automation in the last years. Although these operations are characterized by high repetitiveness, they require an advanced ability to adapt to different scenarios, As such, viable finishing robots are commonly equipped with advanced 3D scanning cameras to detect their surrounding and execute their actions accordingly. A finishing robot is displayed in Figure 3.6 [11].

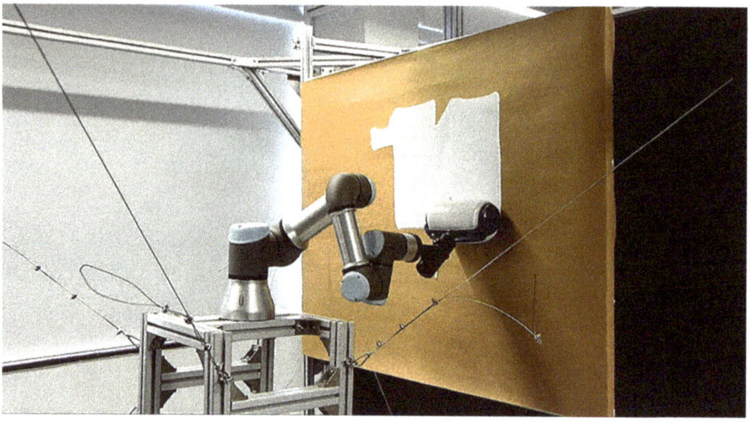

Figure 3.6 A finishing robot [11].

3.4 APPLICATIONS OF CONSTRUCTION ROBOTS

Robotic solutions within construction can automate heavy equipment and fleets for excavation, transportation, load lifting, concrete work, and demolition. The applications of robotics in construction are vast. Common areas of applications of robotics in construction include the following [12]:

- *Automation:* The construction industry stands at the threshold of a remarkable revolution, fueled by rapid advancements in automation and robotics. The automation of construction processes refers to the use of digital manufacturing techniques, including formative, subtractive, and additive methods, and of robots and autonomous or teleoperated vehicles to automate specific tasks or processes in both on-site and off-site fabrication activities. Physically demanding, repetitive installation tasks that are suitable for robotic automation include overhead drilling, electrical, and plumbing fixtures. A concrete construction operation that lends itself to robotic automation is positioning and tying of reinforcing bars, which is characterized by repetitiveness and labor strain and is usually prone to accidents as workers walk on irregular mats of rebar. Figure 3.7 shows robotic automation [13].

Figure 3.7 Robotic automation [13].

- *Paving:* This involves covering excavated land with various materials in different areas of construction sites. Robotics play a key role in this construction phase, with common uses including data identification through radars, area reading, and 3D modeling. Some robots are even designed specifically for automatically paving roads.

- *Project Management:* Drones can play a pivotal role in project management. Numerous companies already leverage drones for diverse tasks, spanning from 3D modeling that provides insights into project landscapes to advanced spatial detection capabilities, drones are reshaping the project management landscape.

- *Demolition:* Using robots to handle large-scale demolitions can make the process faster, safer, and more efficient. Operators can also control the demolition from a safe distance. Demolition robots allow job managers to reduce time and costs because they are able to dismantle structures in record time.

- *Bricklaying:* One of the most popular applications of

construction robots is bricklaying. Bricklaying, either for load bearing and nonload bearing walls or for brickwork facades, is a construction activity that is extremely repetitive and physically taxing, as it implies frequent pick-up of heavy objects and may require labor to work on insecure ladders or scaffolding for long stretches of time. Bricklaying robots are highly skilled at stacking bricks or blocks to construct walls, facades, and other structural parts. Figure 3.8 shows a robot performing bricklaying [14].

Figure 3.8 A robot performing bricklaying [14].

- *Welding:* Some robots are used for various welding jobs. Polar robots, in particular, are great for creating strong and reliable connections between metal parts, such as steel beams. Figure 3.9 depicts a welding robot [11].

Figure 3.9 A welding robot [11].

- *3D printing:* In recent years, additive manufacturing (AM), also known as 3D printing, has grown rapidly in the construction industry. This refers to creating 3D objects by depositing materials using a computer-controlled process. What used to take months to build with traditional methods can now be done in a day or two using 3D printing technology. Some construction robots now have this capability, creating intricate shapes and structures layer by layer using materials like concrete or plastic by following digital designs. AM has been adopted in the construction sector to enable greater flexibility in the geometry of structural elements, reduce material consumption and wastage, and improve worker safety.

- *Site Inspection:* Construction robots, like drones, are equipped with cameras, sensors, and measuring devices that can check structures and keep an eye on construction progress. By using these robots, contractors can quickly identify problems or areas that need attention.

3.5 BENEFITS

Construction has always been known for being physically demanding and time-consuming. But due to the integration of robotics in the construction industry, things are changing for the better. Construction robots have the potential to significantly revolutionize the industry. They are here to speed up tasks and reduce human errors and fatigue-related setbacks. Since construction is one of the most dangerous jobs, robots can eradicate the need for laborers in unsafe and hazardous conditions. Here are some other benefits of robotics in construction [7]:

- *Increased Efficiency:* Robots and automated machinery do not experience human symptoms such as fatigue, burnout, or a lack of interest in a repeated process. Therefore, robots can be more efficient in handling repetitive or mundane tasks, allowing human workers to concentrate on more essential and complex jobs. This translates into cost savings and ensures that projects are completed on schedule.

- *Lower Costs:* In construction, labor accounts for 25-30% of total expenses. However, construction robotics play a significant role in lowering these costs by automating labor-intensive tasks. It is well known that many construction projects overshoot their budgets and timelines because of unforeseen challenges. Construction robots can help in this case by minimizing the need for rework and human errors. More work in less time calls for less operational costs and associated expenses. Since robots can operate for much longer and at higher levels of precision than human workers, this benefit is intrinsic.

- *Improved Safety:* Safety is essential in construction, but accidents can still happen even with extensive training and precautions. Robots can take over dangerous jobs, removing or reducing the risk of injury for humans. A more controlled environment means less injuries at a jobsite. The construction industry has long been one of the most dangerous due to the nature of the job: heavy lifting, hazardous elements, and somewhat perilous terrains or platforms. The implementation of construction robots can result in a significant reduction of injuries on construction sites.

- *Better Sustainability:* Robotic innovations have helped to usher in an era of sustainable building practices, significantly reducing waste and the environmental impact of construction processes. Robots can optimize resources by accurately measuring and cutting materials, recovering recyclable or valuable items, and extending material lifespan and utility, thus promoting a circular economy. The automation capabilities of robots can lead to faster project completion and lower energy consumption, thereby minimizing the environmental carbon footprint of construction activities.

- *Overcoming Labor Shortages:* The biggest challenge the construction industry faces is a shortage of skilled workers. However, introducing robots can help ensure that work progresses steadily, especially for repetitive tasks.

- *Better Quality Control:* The precision of construction robots ensures that every task, from bricklaying to welding, is done with consistent accuracy. This consistency leads to higher-quality work and fewer defects. It also allows leaders to spot and address quality issues on the spot, leading to better outcomes and higher client satisfaction.

- *Increased Productivity:* Robots in construction significantly boost productivity. They can work consistently without getting tired, allowing more work to be completed in a shorter period. The precision of robots ensures high-quality work, reducing the need for rework and further enhancing overall productivity.

- *Enhanced Accuracy:* Compared to human labor, robots are highly accurate. The incorporation of robotics technology eliminates the probability of human errors, delivering highly accurate results. Robots, with their efficiency and accuracy, are at the forefront of the shift towards sustainable construction.

- *Return on Investment:* While the initial costs associated with robotic integration in construction are substantial, the return on investment (ROI) over time can be equally impressive. Robots, with their ability to work round the clock, can drastically reduce project timelines, ensuring faster completions. The precision and consistency brought about by robots mean reduced wastage, leading to cost savings in terms of materials. When these benefits are quantified over the lifespan of the robots, the ROI becomes evident.

Some of these benefits are illustrated in Figure 3.10 [10].

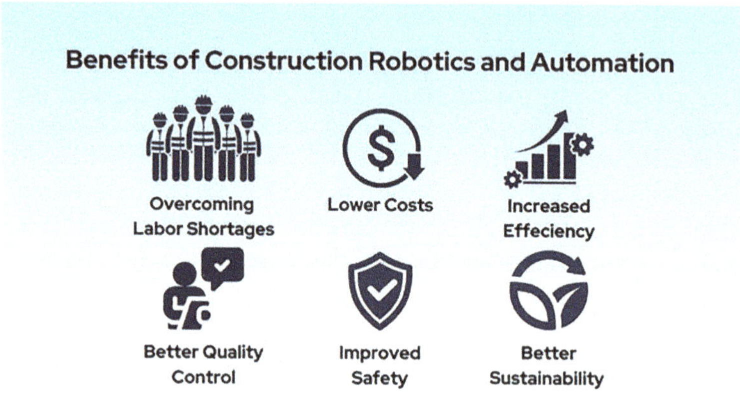

Figure 3.10 Some benefits of construction robotics [10].

3.6 CHALLENGES

While promising a plethora of benefits, robotics also brings with it a set of challenges and concerns. The construction project's requires people of different skills, equipment, materials, and machinery for every activity as result of which the construction process becomes complicated. Humans are able to surpass robots when dealing with decision-making in an unfamiliar situation. Urbanization, coupled with population growth, has led to a surge in infrastructure demands. Meeting this demand requires efficiency and precision, something that traditional construction practices often struggle with. Other challenges include the following [13]:

- *Resistance to Change:* The construction industry has traditionally relied on manual labor, which can cause resistance to adopting new technologies. Skepticism, fear of job displacement, and lack of awareness about the potential benefits of automation may hinder the acceptance and implementation of automated solutions. Automation is coming, whether companies like it or not, and success will depend on recognizing and overcoming this resistance.

- *Labor Shortage:* Basic knowledge of construction software and hardware are becoming increasingly prevalent on construction sites, establishing a new baseline of expectations for skilled workers. According to the Association of General Contractors of America, 80% of contractors have difficulties finding enough skilled workers to satisfy demand. As a labor-intensive industry, construction is the ideal field for a robotic revolution to take place.

- *Safety Concerns:* Safety, or the lack thereof, has been a long-standing issue in the construction industry. Construction sites, with their heavy machinery, high structures, and complex operations, are inherently risky. Integrating robots leads to substantial improvements in site safety. Construction robotics can offer real-time monitoring and safety alerts, as well as enhance site visualization for construction team members.

- *Lack of Flexibility:* Construction projects often involve unique and customized tasks that require flexibility, as well as collaboration and teamwork among various trades and skilled workers. Some construction professionals fear that automation cannot achieve the level of flexibility and improvisation required in complex construction environments. However, the opposite is true. By automating layout, construction professionals can be even more flexible with what is included in layout.

- *Collaboration:* There is a need to work closely with automated solution providers who have expertise in the construction industry. These partnerships ensure that the adopted automated systems are suitable for construction tasks and can effectively address the specific challenges faced by the industry.

- *Fear of Unemployment:* The integration of robotics in the construction industry, while promising a plethora of benefits, also brings with it societal concerns, primarily related to job displacements. There is a fear of unemployment among laborers. Robotics and automation have started replacing a notable amount of the workforce in the construction business, eliminating the manual processes.

- *Maintenance:* Like any other technology, robotics requires maintenance to maintain its efficiency. Maintaining several robots can result in higher expenses.

- *Project Complexity:* The modern world demands infrastructure that is not just functional but also sustainable, aesthetically pleasing, and equipped with the latest technologies. This demand has led to an increase in project complexity. This complexity is not limited to buildings. Mistakes or misalignments, which might have been minor issues in simpler projects, can lead to significant complications, necessitating expensive rework.

3.7 CONCLUSION

The construction industry is undergoing a robotic overhaul. Robotics is giving the construction industry the massive potential to reach its prime. With an incredible amount of potential for productivity, safety and efficiency, autonomous robots are changing construction. The ability to work 24/7 and error-free make robots more appealing than human workers for the future. Unlike humans, robots do not feel fatigue and are able to perform the same task with high precision and consistency. Though the construction industry is highly unautomated, several different robots are already on the market for assisting construction professionals.

Robotics is poised to redefine the construction landscape, making it more efficient, precise, and sustainable. Robotics for construction are only going to be getting better. In the coming decade, we will be able to witness several different inventions in the field of construction automation and robotics that will completely revolutionize the construction business. Robots are important part of the future of construction. More information on robotics in the construction industry can be found in the books in [15-18] and the following related journals:

- *Robotica*
- *Robotics*
- *Robitics and Autonomous*
- *Robotics and Computer-Integrated Manufacturing,*
- *Advanced Robotics*

- *Autonomous Robots*
- *Automation in Construction*
- *Journal of Robotics*
- *Journal of Robotic Systems*
- *Journal of Robotic Surgery*
- *Journal of Robotics and Mechatronics*
- *Journal of Intelligent & Robotic Systems*
- *Journal of Mechanisms and Robotics-Transactions of the ASME*
- *Journal of Automation, Mobile Robotics and Intelligent Systems*
- *Journal of Future Robot Life*
- *IEEE Robotics and Automation Letters*
- *IEEE Transactions on Robotics*
- *International Journal of Medical Robotics and Computer Assisted Surgery*
- *International Journal of Robotics Research*
- *International Journal of Social Robotics*
- *International Journal of Humanoid Robotics*
- *International Journal of Advanced Robotic Systems*
- *Science Robotics*
- *Soft Robotics*

REFERENCE

[1] “Transforming the future of construction,” Unknown Source.

[2] “Use of robotics in construction: Past, present, and future trends,

https://www.hp.com/us-en/printers/site-print/use-robotics-in-construction-past-present-future-trends.html

[3] M. N. O. Sadiku, P. A. Adekunte, and J. O. Sadiku, “Robotics in the construction industry,” *International Journal of Trend in Scientific Research and Development*, vol. 9, no. 1, January-February 2025, pp. 84-93.

[4] “Human–robot interaction,” *Wikipedia*, the free encyclopedia

https://en.wikipedia.org/wiki/Human–robot_interaction

[5] “Robotics,” *Wikipedia*, the free encyclopedia

https://en.wikipedia.org/wiki/Robotics

[6] M. N. O. Sadiku, U. C. Chukwu. A. Ajayi-Majebi, and S. M. Musa, “A primer on robotics, ” *International Journal of Trend in Scientific Research and Development*, vol. 6, no. 7, November-December 2022, pp. 614-621.

[7] “Different types of robots transforming the construction industry,” October 2020,

https://roboticsbiz.com/different-types-of-robots-transforming-the-construction-industry/

[8] “How construction robotics is going to change the industry forever,” February 2024,

https://www.cemexventures.com/how-construction-robotics-is-going-to-change-the-industry-forever/

[9] “Construction robot,” Unknown Source.

[10] E. Thompson, “How robots are transforming the construction

industry," August 2024,

https://www.cyngn.com/blog/how-robots-are-transforming-the-construction-industry

[11] "A fully automated construction industry? Still a long road ahead,"

https://www.archdaily.com/945761/a-fully-automated-construction-industry-still-a-long-road-ahead

[12] "The power of robotics in construction," September 18, 2024

https://standardbots.com/blog/revolutionizing-construction-the-power-of-robotics?srsltid=AfmBOoqlPvNWkc1N2bheLVJfVGVdT8JEYJxmEjcGEQBY1etR48Y_upJt

[13] "Building the future with robotics in the construction industry,"

https://www.forbes.com/sites/ronschmelzer/2024/10/18/building-the-future-how-ai-is-revolutionizing-construction/

[14] G. Baleri, "2024 is the year for autonomous robots in construction," January 2024,

https://www.constructionexec.com/article/2024-is-the-year-for-autonomous-robots-in-construction

[15] M. N. O. Sadiku, *Robotics and Its Applications*. Moldova, Europe: Lambert Academic Publishing, 2023.

[16] H. Jebelli et al. (eds.), *Automation and Robotics in the Architecture, Engineering, and Construction Industry*. Springer, 2022.

[17] T. Bock and T. Linner, *Robot-Oriented Design: Design and Management Tools for the Deployment of Automation and Robotics in Construction*. Cambridge University Press, 2015.

[18] A. Chamberlain, *Automation and Robotics in Construction XI*. Elsevier Science, 2012.

CHAPTER 4
DRONES IN CONSTRUCTION

"I think the greatest threat to the privacy of Americans is the drone, and the use of the drone and the very few regulations that are on it today, and the booming industry of commercial drones."

—Dianne Feinstein

4.1 INTRODUCTION

Drone photos, videos, and imagery are used to scope out projects, track building progress, and provide real-time updates. With their real-time data recording and unique aerial advantage, drones can improve efficiency, cut costs, and streamline workflow. Drones are equipped with multiple features to help capture key data on building sites, including cameras, GPS units, thermal sensors, and infrared sensors.

Drones have quickly become an effective tool for businesses of all sizes. Although many industries are regular users of drone technology, the construction industry has both the largest market share and the fastest-growing market share for drone usage over the past few years. These flying vehicles typically have cameras in them, which means that they provide a birds-eye view of a construction site. With sensors and cameras, drones can also track construction equipment and vehicles on a site.

Drones offer immense possibilities and applications in the world of construction, offering benefits ranging from on-site safety to remote monitoring. They have been making a positive impact on the construction world since becoming a part of the industry years ago. Their current capabilities allow them to cut costs, time, risk, and labor, while improving workflow, accuracy, communication, and efficiency. Whether drones are being used to conduct land surveys or to keep track of equipment, drones are proving invaluable to construction [1].

This chapter explores various ways of using drones in the construction industry. It begins with explaining what a drone is. It presents some applications of drones in construction. It highlights the benefits and challenges of drones in the construction industry. It concludes with comments.

4.2 WHAT IS A DRONE?

The FAA defines drones, also known as unmanned aerial vehicles (UAVs), as any aircraft system without a flight crew onboard. Drones include flying, floating, and other devices, including unmanned aerial vehicles (UAVs), that can fly independently along set routes using an onboard computer or follow commands transmitted remotely by a pilot on the ground. A typical drone is shown in Figure 4.1 [2]. A drone is usually controlled remotely by a human pilot on the ground, as typically shown in Figure 4.2 [2].

Figure 4.1 A typical drone [2].

Figure 4.2 A drone is usually controlled by operators on the ground [2].

Drones can range in size from large military drones to smaller drones. Drones, previously used for military purposes, have started to be used for civilian purposes since the 2000s. Since then, drones have continued to be used in intelligence, aerial surveillance, search and rescue, reconnaissance, and offensive missions as part of the military Internet of things (IoT). Today, drones are used for different purposes such as aerial photography, surveillance, agriculture, entertainment, healthcare, transportation, law enforcement, etc.

Drones work much like other modes of air transportation, such as helicopters and airplanes. When the engine is turned on, it starts up, and the propellers rotate to enable flight. The motors spin the propellers and the propellers push against the air molecules downward, which pulls the drone upwards. Once the drone is flying, it is able to move forward, back, left, and right by spinning each of the propellers at a different speed. Then, the pilot uses the remote control to direct its flight from the ground [3].

Drone laws exist to ensure a high level of safety in the skies, especially near sensitive areas like airports. They also aim to address privacy concerns that arise when camera drones fly in residential areas. These include the requirement to keep your drone within sight at all times when airborne. In the United States, drones weighing less than 250g are exempt from registration with civil aviation authorities. If your drone exceeds 250g in weight, you will also require a Flyer ID, which requires passing a test [4]. It is necessary to register as an operator, be trained as a pilot, and have civil liability insurance, in addition to complying with various flight regulations, and those of the places where their use is permitted.

Most drones have a limited payload, usually under 11 pounds. Drones are classified according to their size. Here are the different drone types:

- Nano Drone: 80-100 mm
- Micro Drone: 100-150 mm
- Small Drone: 150-250 mm
- Medium Drone: 250-400 mm
- Large Drone: 400+ mm

One of the emerging trends in drone use for factories is the utilization of LiDAR technology. LiDAR stands for Light Detection and Ranging. This technology provides accurate depth information essential for understanding the three-dimensional structure of the environment. LiDAR sensors emit laser beams to measure distances to objects, creating high-resolution 3D maps of the surrounding terrain and objects. The ability to capture detailed data through LiDAR technology has opened up opportunities for better predictive maintenance, reduction in inspection times, and overall cost savings [5].

4.4 APPLICATIONS OF DRONES IN CONSTRUCTION

A drone in the manufacturing environment is shown in Figure 4.3 [6]. While there are many types of drones, commercial drones are the standard commonly used in construction. Drones can be used for a variety of different things throughout the construction industry. They provide construction teams with an overhead view of jobsites, materials, machinery, and people Common applications of drones in construction include the following [7]:

Figure 4.3 A drone in the construction environment [6].

- *Mapping and Land Surveys:* All drone types offer high-quality cameras capable of achieving survey-grade results, with high accuracy. Consulting topographic maps is essential when planning complex, large-scale construction projects. Although topographic maps are helpful, they are expensive and time-consuming to produce. Drones generate topographic map (that shows changes in elevation) with the help of LiDAR lasers. Due to their ability to map vast quantities of land, drones can exponentially cut down on time spent visualizing a site's topography. This allows the construction crew to pinpoint challenges during pre-construction and spot mistakes, saving time and money. A typical land survey is shown in Figure 4.4 [8].

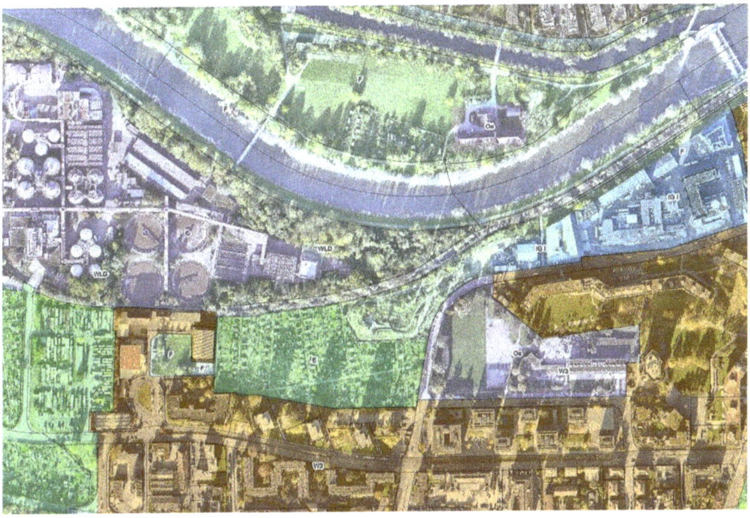

Figure 4.4 .

- *Equipment Tracking:* Losing track of where equipment is located at each job site is a common problem. Equipment security is a very important component of managing a job. A manager or drone operator can conduct a flyover to quickly see if a piece of equipment is in a secure enough location and ensure that the equipment is where it should be.

- *Worker Safety:* Worker safety is a top priority of most construction companies. Workers often have to navigate around hazardous conditions when taking manual measurements. Drones can replace workers in these situations and mitigate the risk construction workers face on the field. Construction managers can also use drone video cameras to monitor the job site for safety concerns and ensure that no equipment are loose or unstable. If there is any suspicious activity, drones can send an alert to management to help prevent vandalism or equipment theft.

- *Road Construction:* Building a road is an uphill task which involves a lot of complex engineering and precision work. Every stage of the construction requires careful planning and attention to detail. Drones are making the process a lot easier by providing geographic information system (GIS) data on the road that civil engineers and construction crews can rely on to make more strategic decisions. They also help direct the pouring of asphalt and concrete and provide detailed visual insight throughout the project. They are a powerful tool for assessing the soil before road construction begins. Figure 4.5 shows the use of a drone in road construction [9].

Figure 4.5 Using a drone in road construction [9].

- *Roof Inspections:* This benefits immensely from using drones. Construction drone photography can eliminate the need for manual roof inspections, which can involve roofs that are wet, slippery, or in dangerously unstable conditions. Drones can get to construction spots, like parts of rooftops that otherwise would be hard and potentially risky to reach. Drones in construction safety can make a big difference in protecting workers.

4.5 BENEFITS

Drones in the construction industry have become commonplace, and the benefits of drones in construction are many. All the drone applications help you avoid making costly mistakes. The tiniest miscalculation spread across miles of road can turn into a big problem. The amount of time and money that drones in construction save is largely substantial. Implementing drones in a construction workflow can completely change the companies' operations throughout the entire bidding, planning, and building process. Through the surveying process, drones can save time and manpower while also mitigating human error. Other benefits of drones in construction include the following:

- *Speed:* One of the main benefits of using drones in construction is their speed. Drones can collect data faster because they are not slowed down by on-the-ground hazards. In the end, road construction surveys that used to take several weeks can now be done in a matter of days with drones.

- *Security:* With large construction sites, the issues of theft, trespassing, and vandalism make monitoring construction sites difficult to manage. To combat this, construction companies use drones to surveillance the area. Along with monitoring construction project progress, drones can do around-the-clock, real-time job site monitoring. They can protect the site from intruders like unaccompanied visitors or even thieves or vandals.

- *Collaboration:* Drones have allowed construction companies to improve communication and collaboration across their workforce. They can improve internal collaboration for teams by sending information to connected software during flyovers. Design teams, engineers, construction managers, workers, and owners can access the data simultaneously, follow the project, and catch any mistakes that may have occurred.

- *Improved Decision Making:* Drone data helps improve on-the-ground decision-making. A drone can also give decision-makers regular updates and keep everyone informed about a construction's progress. With these insights, those in command of a site can catch pitfalls and mistakes before they balloon into a larger problem.

- *Improved Worker Safety:* Drones increase worker and site safety. Since drones capture data from the air, you do not need to impose unnecessary risk on workers when inspecting the road in or near moving traffic. Overall, drones lower worker and site safety issues to a much more manageable level. This leads to fewer worker accidents, less liability, and lower insurance premiums.

- *Reduced Labor:* Adapting drones into the construction industry lessens the amount of labor and time needed to create accurate surveys. Drones also lessen the risk of human errors and capture crucial data in a fraction of the time as traditional methods. With the help of drones, the project manager can monitor overall productivity, enabling him to keep workers productive.

- *Cost Saving:* For any construction company, there are three main aspects that matter – time, quality and cost. Drones are more affordable than mobilizing a technical inspection team. A construction company's investment in construction drone services is a practical and future-forward decision. The return on investment (ROI) speaks for itself through many business benefits such as increased profit margins, more effective customer acquisition, and staying on schedule.

- *Efficiency:* Anyone who has ever been through the process of manually surveying a construction site can tell you that it is not an easy process. It takes time, a workforce with a very particular skill set, money, and a willingness to redo the survey if anything goes wrong. With drone surveying, even beginner surveyors are capable of surveying 120 acres per hour for a 60x improvement in surveying efficiency.

- *Saving Time:* Drones can help avoid material shortages and overages by calculating stockpile volumes and tracking changes to earthwork. They can also monitor sites for changes that might indicate potential hazards, such as landslips.

4.6 CHALLENGES

Any technology tool is only as good as your ability to use it, and drones are no exception. Construction sites do not exist in vacuums; they operate on deadlines and are subject to weather and are themselves rugged environments. Other challenges of drones in construction include the following:

- *Training and Certification:* Drones may be small, but they are still a type of aircraft. They require training and certification to operate. A nation's aviation authority regulates all unmanned aerial vehicles (or UAVs). The location of the construction worksite can determine how, when, and if you fly, depending on the airspace overhead.

- *Regulations:* Flying drones is highly regulated. If you are close to an airport or other restricted airspace, you may find that you cannot legally fly a drone. Most countries have different sets of regulations for drones. US commercial drone regulations are managed by the FAA.

- *Injury:* When drones veer off course, they can hit people and their sharp blades can cause injury and death. Construction projects can be dangerous. Construction worker fatalities in 2019 represented over 20% of all on-the-job deaths, making construction one of America's most dangerous industries. A significant portion of construction project costs are injury related. This leads to increased insurance prices and reduced profit margins. Injury may cost lost project time, litigation, medical bills, and compensation. A deadly accident can end a construction company all together.

- *Hacking:* Like most connected devices, drones are vulnerable to hacking. Hackers can quickly attack the central control systems of some commercial drones from a mile away to access private information, destroy or damage files, and leak data to unauthorized third parties.

4.7 CONCLUSION

The use of drones in construction and construction-related industries has now become commonplace, with construction being the fastest-growing commercial industry adopting drones. Drones have become the go-to tool for construction firms to track, map, survey, inspect, and manage worksites more efficiently and safely. They are increasingly becoming pivotal in the construction industry, bringing unparalleled value in terms of efficiency, safety, and accuracy. For drone pilots looking to launch a new career in commercial drone flight, the construction industry is a field worth consideration.

The impact of drones is so great that companies will fall behind industry standards without them. While the impact of drones in the construction industry is evolving, they are already revolutionizing the sector and changing the way construction projects are handled. As technology advances, drones will be able to take on more construction tasks than ever before. Drones will continue to offer industry leading solutions that deliver the highest quality of speed, accuracy, and cost effectiveness. It is safe to say that the commercial drone is here to stay. More information about drones in construction can be found in the books in [10-13].

REFERENCE

[1] M. N. O. Sadiku, M. Oteniya, J. O. Sadiku, and S. Abunene, "Drones in construction," *International Journal of Scientific and Research Publications*, vol. 14, no. 9, September 2024, pp. 285-294.

[2] "Drone in construction & infrastructure,"

https://www.jouav.com/industry/drone-in-construction

[3] "How drones work and how to fly them," May 2024,

https://dronelaunchacademy.com/resources/how-do-drones-work/

[4] "What are the main applications of drones?" June 2024,

https://www.jouav.com/blog/applications-of-drones.html

[5] "Drones in manufacturing: A game-changer for industry,"

https://viper-drones.com/industries/infrastructure-drone-use/manufacturing/#:~:text=The%20integration%20of%20drones%20into,on%20manufacturing%20is%20no%20exception.

[6] J. Moraglia, "Benefits of using a drone for construction," January 2024,

https://thedronelifenj.com/benefits-of-using-a-drone-for-construction/

[7] L. Stannard, "6 Ways drones in construction are changing the industry," February 2022,

https://www.bigrentz.com/blog/drones-construction?srsltid=AfmBOooXvjQzNrIfYOlUib380fStZaXLZCBA8mN8C18PzaYyy79ysTHP

[8] "Six ways drones in construction add value,"

https://wingtra.com/drone-mapping-applications/drones-in-construction-and-infrastructure/

[9] "Drones in construction – Why they are beneficial and how to

use them,"

https://www.propelleraero.com/blog/drones-in-construction-why-they-are-beneficial-and-how-to-use-them/

[10] D. Tal and J. Altschuld, *Drone Technology in Architecture, Engineering and Construction: A Strategic Guide to Unmanned Aerial Vehicle Operation and Implementation*. Wiley, 2021.

[11] F. D'Alessandro, *Drones in Construction Sites*. CreateSpace Independent Publishing Platform, 2015.

[12] L. P. Mamani, *Making Money with Drones, Drones in the Construction Industry*. Independently Published, 2021.

[13] J. Powell, *Drone Services in Construction*. Office of Industries, U.S. International Trade Commission, 2021.

CHAPTER 5
INTERNET OF THINGS IN THE CONSTRUCTION

"The Internet of things in construction is not just about tracking equipment; it's about using real-time data to optimize processes, improve safety, and build smarter, more efficient structures."

—Anonymous

5.1 INTRODUCTION

Digital transformation is the process of integrating technology to improve business processes. It allows for the automation of time-consuming tasks. Technology has come a long way when we refer to the construction industry. In the construction industry, IoT devices can help with digital transformation by allowing software construction companies to access real-time data. The industry is currently undergoing critical transformation that will improve efficiency, well-being, process improvement, and the introduction of new tools. It is poised for digital transformation, and companies that fail to adapt quickly risk being left behind.

The construction industry, also known as the building industry, is one of the oldest businesses that humans developed. All other industries rely on the construction to create their infrastructure and architecture. According to the Occupational Safety and Health Administration (OSHA), a division of the US Department of Labor, there are approximately 252,000 construction sites in the United States that employ around 6.5 million people. Construction serves as a base for urbanization and also for employment and generation of new complexes and buildings. With the rise of urbanization and population density, the construction sector has become one of the most important contributors to global prosperity. Construction efficiency and profitability have a direct influence on the global economy. Figure 5.1 shows a typical construction site [1], while Figure 5.2 shows a typical construction worker [2].

Figure 5.1 A typical construction site [1].

Figure 5.2 A construction worker [2].

However, the construction industry has been relatively slow in adopting and capitalizing on new technologies like IoT. Historically, construction has not been the most tech-centric industry, but that is changing. Project managers and investors are still wary about implementing IoT technology; some do not know how to use it, others doubt its efficiency. Today, the industry is embracing IoT technology to enhance productivity, optimize operational efficiency, and improve on-site safety.

The modernization of the construction industry is one of the major opportunities for investors, innovators, thought leaders, and professionals. The Internet of things (IoT) is a system of interrelated smart devices that are fitted with sensors, software, and other technologies that facilitates communication. IoT is revolutionizing industries worldwide, and construction is no exception. IoT technology in construction is improving safety, efficiency, and project management, revolutionizing construction operations and driving innovation. It will have a part to play in waste management and employee safety improvement, construction costs reduction, and helping match properties against the requirements of prospective tenants. Contractors and builders are leveraging the transformative power of IoT in construction to ensure maximized client satisfaction and minimized costs [3].

In this chapter, we will explore how the construction industry is embracing the Internet of things (IoT) as a transformative tool. It begins with providing an overview on what Internet of thing is all about. It describes industrial IoT. It explains IoT in construction and presents some of its applications. It highlights the benefits and challenges of IoT in the construction industry. It concludes with comments.

5.2 OVERVIEW ON INTERNET OF THINGS

The concept of the Internet of things (IoT) has been around since the late 1990s, but it gained momentum in the 2000s with the rise of Internet-connected devices. The Internet began with some military computers in the Pentagon called Arpanet in 1969. It expanded throughout the 1980s as a set of four parallel military networks, each at a different security level. The core technology which gives the Internet its particular characteristics is called Transmission Control Protocol/Internet Protocol (TCP/IP), which is essentially a set of rules for communication [4].

Internet of things (IoT) is a worldwide network that connects devices to the Internet and to each other using wireless technology. IoT is expanding rapidly and it has been estimated that 50 billion devices will be connected to the Internet by 2020. These include smart phones, tablets, desktop computers, autonomous vehicles, refrigerators, toasters, thermostats, cameras, alarm systems, home appliances, insulin pumps, industrial machines, intelligent wheelchairs, wireless sensors, mobile robots, etc. Figure 5.3 illustrates the Internet of things [5].

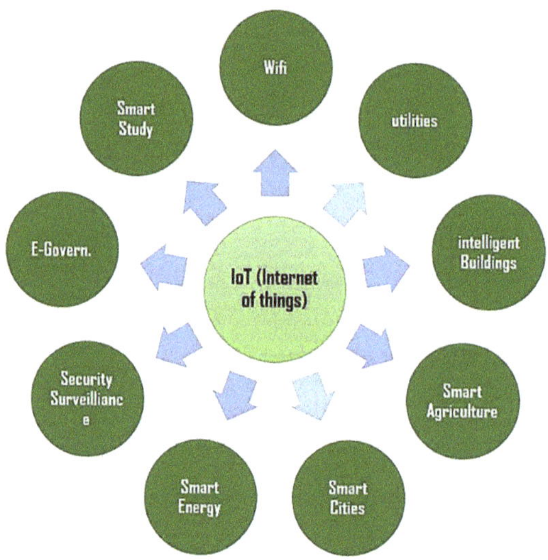

Figure 5.3 The Internet of things [5].

There are four main technologies that enable IoT [6]: (1) Radio-frequency identification (RFID) and near-field communication, (2) Optical tags and quick response codes: This is used for low cost tagging, (3) Bluetooth low energy (BLE), (4) Wireless sensor network: They are usually connected as wireless sensor networks to monitor physical properties in specific environments. Communications technologies in Internet of things are portrayed in Figure 5.4 [7].

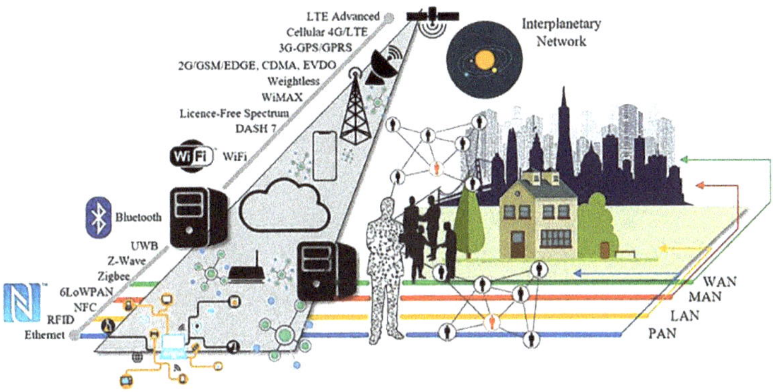

Figure 5.4 Communications technologies in Internet of things [7].

IoT technology enables people and objects to interact with each other. It is employed in many areas such as smart transportation, smart cities, smart energy, emergency services, healthcare, data security, industrial control, logistics, retails, structural health, traffic congestion, manufacturing, and waste management. The Internet of things is extensively developed world-wide with a focus on civilian applications such as electric power distribution, intelligent transportation, healthcare, industrial control, precision agriculture, environmental monitoring, etc. Figure 5.5 shows a typical representation of IoT [8].

Figure 5.5 A typical representation of IoT [8].

5.3 INDUSTRIAL INTERNET OF THINGS

The growth of the Internet of things (IoT) is drastically making impact on home and industry. While the IoT affects among others transportation, healthcare, or smart homes, the Industrial Internet of Things (IIoT) refers in particular to industrial environments. IIoT is a new industrial ecosystem that combines intelligent and autonomous machines, advanced predictive analytics, and machine-human collaboration to improve productivity, efficiency and reliability. It is bringing about a world where smart, connected embedded systems and products operate as part of larger systems [9].

The industrial Internet of things (IIoT) refers to the application of the Internet of things (IoT) across several industries such as manufacturing, logistics, oil and gas, transportation, energy/utilities, chemical, aviation and other industrial sectors. A typical industrial Internet of things is shown in Figure 5.6 [10].

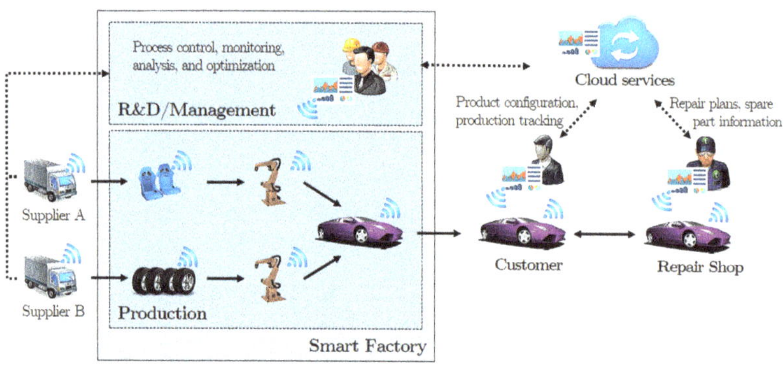

Figure 5.6 A typical industrial Internet of things [10].

IIoT is often used in the context of Industry 4.0, the Industrial Internet and related initiatives across the globe. Industry 4.0 describes a new industrial revolution with a focus on automation, innovation, data, cyber-physical systems, processes, and people [11]. With Industry 4.0, the fourth industrial revolution is set on merging automation and information domains into the industrial Internet of things, services, and people. The communication infrastructure of Industry 4.0 allows devices to be accessible in barrier-free manner in the industrial Internet of things, without sacrificing the integrity of safety and security [12].

5.4 IOT IN CONSTRUCTION

The Internet of things (IoT) relies on connected sensors and embedded systems to collect data and control devices remotely. Connected sensors can tell the location of people or equipment, monitor security, collect data on job site conditions, and automate inspections. Cameras placed around the site or mounted on drones can help with monitoring and inspections, even if the inspector is in another location. For example, workers in the construction industry can use GPS devices mounted on equipment. These can communicate the equipment's real-time location to construction managers, helping them streamline resource management.

The term IoT in construction refers to the use of technological equipment or Internet of things and modern-day Internet software in the process of construction to be able to maximize the efficacy of the project. The IoT in construction market is categorized into hardware, software, connectivity, and services. The market is classified as residential and non-residential. It can also be classified into asset monitoring, predictive monitoring, fleet management, wearables, and others. The market forecast is analyzed across North America, Europe, Asia-Pacific, and LAMEA. China is the largest shareholder of the IoT in construction industry, owing to the large manufacturing and connectivity infrastructure and a strong foothold in the construction industry [13]. Figure 5.7 shows the symbol of IoT in construction [14].

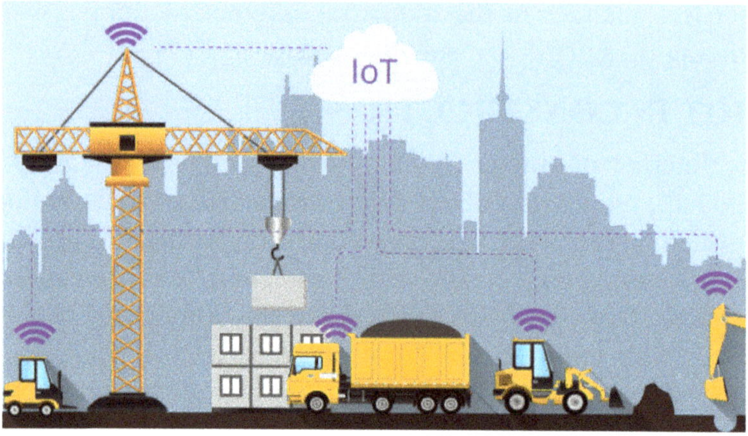

Figure 5.7 The symbol of IoT in construction [14].

A wide range of IoT devices are important in the construction industry. IoT devices include the following types:

- *Smart Sensors:* These are embedded in construction equipment and structures to monitor various parameters such as temperature, humidity, vibration, and structural integrity.
- *Wearable Devices:* Wearable devices, such as smart helmets, vests, and glasses, enhance worker safety and productivity.
- *Drones:* Drones equipped with cameras and sensors are used for site surveys, progress monitoring, and inspections. They provide high-resolution images and 3D models, enabling better planning and decision-making.
- *Connected Machinery:* IoT-enabled machinery and equipment can communicate with each other and central management systems. This connectivity allows for real-time monitoring, remote control, and predictive maintenance, reducing downtime and improving efficiency.

These IoT devices enable construction managers to monitor construction sites more effectively in various different ways. These IoT devices are available to optimize the process on a construction site in various different phases. They are key to tracking tools, equipment, and machines on construction sites. Here are some common IoT sensors and controllers in the industry [15]:

- *Wearable sensors* allow managers to track workers for safety and productivity.
- *GPS trackers* communicate the location of equipment and vehicles.
- *Embedded sensors* can track equipment performance for efficiency improvements and predictive maintenance.
- *Materials tracking* is possible with smart labels that provide location and quantity data.
- *Wearable sensors* allow managers to track workers for safety and productivity.
- *GPS trackers* communicate the location of equipment and vehicles.
- *Embedded sensors* can track equipment performance for efficiency improvements and predictive maintenance.

5.5 APPLICATIONS OF IOT IN CONSTRUCTION

The spectrum of IoT applications is vast, comprising wearables, traffic monitoring, e-health, agriculture, hospitality, smart grid, smart homes, energy saving, maintenance management, food production, etc. IoT has diverse applications in construction, from wearables for worker safety to sensors for structural health monitoring. Common areas of applications include [14,17,18]:

- *Site Monitoring:* This is a crucial component of IoT application in construction, encompassing the use of sensors to continuously record jobsite conditions. Construction sites are enormous and relying solely on manual security is not feasible or practical. IoT security strategy for a construction site can work on preventing intrusion and quick detection as well as recovery from unwanted activities. IoT enables remote monitoring of environmental factors like temperature, humidity, noise, and vibration, ensuring safety and regulatory compliance. IoT-connected surveillance cameras and various measurement sensors allow controlling and adjusting machinery remotely, increasing the accuracy of operations. All of that helps coordinate building activities more efficiently, reduce idle time, and link workers' performance with relevant tasks and processes on-site.

- *Site Safety:* One of the most primitive and raw issues on a construction site is the safety of the people working there from various disasters. Construction is known for one of the higher fatal injury rates worldwide. Everyone from the government to the owners, managers, and workers, people at every level, are concerned about the safety of people at the construction site. Construction sites frequently have risked such trench collapse, falls, scaffold collapse, insufficient protective equipment, repetitive motion injuries, and others. IoT devices can enhance efficiency and safety. They can make work sites safer by using IoT sensors to monitor access and track workers. The advent of IoT technology in construction has substantially heightened site safety standards. The tracking can allow for immediate response in an emergency. IoT solutions significantly boost on-site safety, integrating innovative technologies for hazard monitoring and worker protection in the construction sector. Smart wearable technologies, like helmets and boots connected to IoT systems, track vital signs and locations, thus alerting workers to imminent dangers and even enabling human resources managers to oversee personnel effectively. Figure 5.8 shows how IoT enhances safety in the construction industry [18].

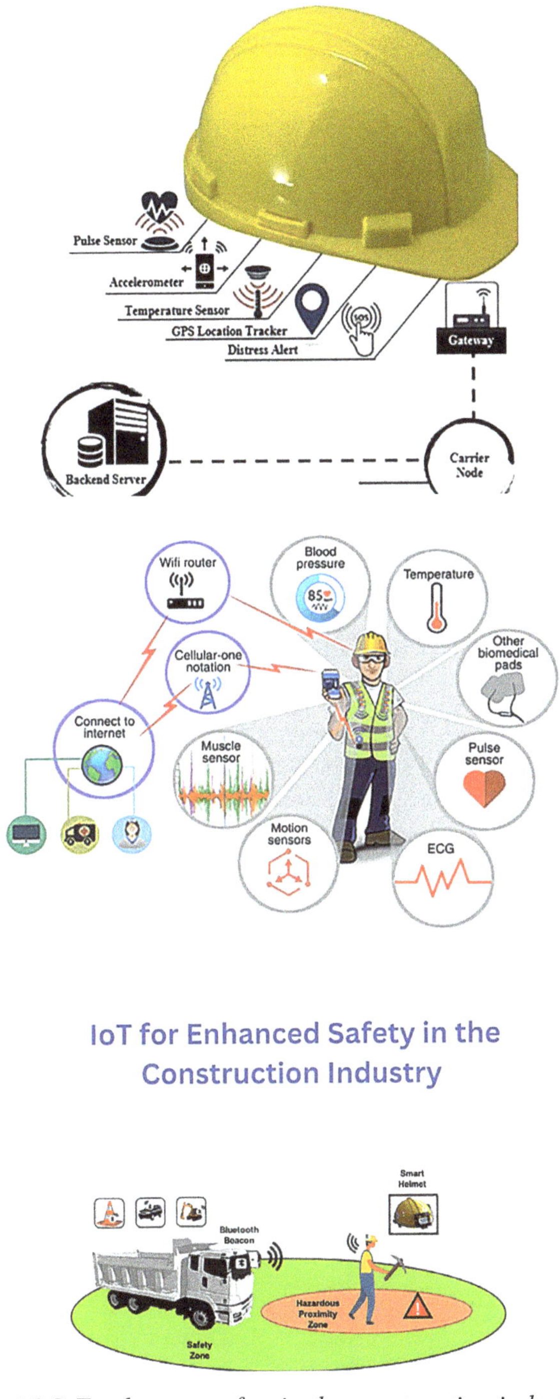

Figure 5.8 IoT enhances safety in the construction industry [18].

• *Concrete Monitoring:* Construction sites can use IoT devices to monitor their materials. Concrete sensors are one of the most prominent examples. Concrete dries at different rates depending on environmental conditions and the type of mixture companies use. If decision-makers do not accurately estimate how fast a pour will dry, they could delay their entire project, potentially leading to lost revenue and unhappy clients. Construction companies can embed wireless sensors at multiple depths before a pour to track the drying rate of concrete. By placing connected sensors in concrete, construction teams can remotely check in on the curing process or see if it needs repair. Project managers can use pre-set triggers to receive an alert once the sensors detect the concrete has met a certain threshold or target. This feature eliminates the need for on-site visits, significantly increasing efficiency.

• *Project Management:* Completing a project within a limited budget is a key performance indicator for project managers as well as construction companies. The use of IoT in construction can help contractors utilize the resources available to them in an efficient way. IoT devices can facilitate cost-cutting by using site monitoring techniques for monitoring vehicles, equipment, material utilization, thereby keeping the project budget-friendly. IoT also helps in completing the project at a faster rate.

• *Fleet Management:* IoT technology has the potential to revolutionize fleet management in construction. The technological advancement in fleet management, powered by IoT, represents a significant shift in how construction companies manage their fleet of vehicles and machinery. It is a move towards more data-driven decision-making, where every aspect of the fleet is monitored and optimized for peak performance. IoT is steadily redefining fleet management, transcending traditional methods by introducing automated tracking of both vehicles and essential equipment. This advancement is pivotal in augmenting productivity, bolstering safety measures, and elevating site monitoring effectiveness.

• *Waste Management:* This not just calls for an environmental effect and degradation but it also calls for a dumping issue that needs to be resolved on-site of the waste and the debris.

Construction business owners are obliged by the law to dispose of debris and recycle waste. Construction teams use IoT technologies to keep track of and minimize their waste. IoT tracking provides the insight the industry needs to reduce waste. Data from vehicle tracking systems can reveal inefficient usage patterns. The used IoT can help create a sustainable plan for using the recycled products, adopting zero waste terminology. Besides having to oversee waste management for the sake of compliance with governmental regulations, disposing of debris regularly allows property owners to create more space on the construction site and reduce the risks of hazards.

- *Machine Control:* This stands as a pivotal aspect of IoT in construction, delivering unparalleled precision in machinery operations and real-time updates on progress, movements, and status. Many heavy machines are involved in construction work and these machines are operated by humans. This task can further be simplified through IoT. Primarily utilized in heavy civil construction, the scope of machine control technology is expanding into diverse projects as the technology evolves. Incorporating IoT in machine control, such as through wearables and safety sensors, exemplifies its unique applications in construction, significantly boosting operational efficiency and on-site safety.

- *Theft Prevention:* This is another growing application of IoT in construction. Construction sites are prime targets for thieves, given their expensive equipment and relatively low security. Now, as construction material costs are rising, this theft could grow, but IoT devices help fight it. IoT trackers can transmit a vehicle or tool's location in real time without alerting the person who took it. Construction companies or police forces can then track down and recover the stolen goods.

- *Wearables:* These are the devices that need to be worn by the people working on a construction site, especially the laborers. This can be used to alarm the worker if he is too close to a dangerous zone. Also, this is used to track the location in the case of a disaster or injury that happens with a worker.

- *Augmented Reality (AR):* This is a very popular application of IoT. The construction industry has been using it for some time now. In AR, 3D models are created and combined with all building data. Users can wear an AR device and then they get to experience it in real-time. The most common applications of AR in the construction industry include project planning, automated measurements, project modification, onsite project information, team collaboration, and safety training. By using this technology, teams can organize presentations, discuss and examine every detail of the project. They can make changes before it becomes expensive to alter. Another vital use of AR is in safety training.

- *Equipment Maintenance:* Without technology, equipment breakdowns are challenging to anticipate. Someone has to manually check oil levels, emissions, and vibration rates manually, and use their expertise to determine how close the machine is to breaking down. With IoT, that process is automatic. IoT sensors can wirelessly measure vibration, pressure, temperature, and moisture in real-time, sending any unusual readings directly to a decision-maker's device.

- *Weather Forecasting:* Project managers have to consider environmental conditions when scheduling subcontractors, pouring cement, and approving the use of some equipment. Unfortunately, the local weather forecast is not always reliable. Fortunately, IoT offers a solution. Project managers can use wireless IoT sensors to track humidity, temperature, precipitation, and wind speed to forecast upcoming weather events.

5.6 BENEFITS

Applying IoT lengthens the life of your machinery and increases its resale value. IoT's integration into construction sites goes beyond mere tracking; it encompasses timely maintenance, efficient management of parts, and optimizes the utilization of machines. Several builders and construction companies are using mobile applications to improve efficiency, enhance safety, access real-time data, increase mobility and save time. With IoT-enabled sensors, you can track the location of the vehicle fleet and position landscaping equipment precisely. Other benefits include the following [15,19]:

- *Smart Construction:* This entails the integration of IoT devices and sensors into building materials, tools, and equipment used on construction sites. By monitoring energy consumption, waste generation, and environmental impact in real-time, construction companies can make informed decisions to minimize their carbon footprint and promote eco-friendly construction methods.

- *Smart Helmets:* One of the remarkable applications of IoT in the construction industry is the development of smart safety helmets. These helmets are integrated with sensors that detect and monitor environmental conditions, such as temperature, humidity, and toxic gases. By continuously collecting data in real-time, these helmets can alert workers and supervisors about potential hazards, enabling them to take immediate preventive actions

- *Sustainable Construction:* IoT plays a vital role in promoting eco-friendly practices. IoT devices can monitor energy consumption, optimize resource usage, and facilitate waste management, contributing to sustainable construction processes. The data collected through IoT can also support the implementation of green building certifications and aid in measuring environmental impact.

- *Fuel Saving:* The cost of each construction project corresponds to the fuel consumption because large sites require using machines that burn thousands of dollars per day. But with the appropriate control over expenditures, effective refueling management, timely maintenance, and the ability to upgrade machinery, you can reduce the project's final cost and its duration dramatically. All of that can be done using IoT-enabled fuel sensors on your construction site.

- *Concrete Curing:* With concrete curing sensors, you can speed up construction schedules. Such IoT-enabled devices provide monitoring of concrete's maturity via temperature probes and then transmit real-time data to the cloud. You have to embed the sensors in concrete during casting for tracking its curing. That allows managers to plan further construction work accurately.

- *Improved Workflow Management:* With real-time data from sensors, you can adjust operations to maximize efficiency and resources. IoT-informed operational optimizations can cut labor costs by up to 14.2%.

- *Data Security:* With the increase in connected devices, data security becomes paramount. Implement robust cybersecurity measures to protect your data and ensure compliance with relevant regulations.

- *Automation:* IoT will drive automation. Automated processes can streamline everything from cost estimation and quantity surveying to labor scheduling and materials ordering. By automating processes and enabling seamless data exchange, IoT devices simplify project management and keep managers abreast of project developments, facilitating informed decisions

- *Inspection:* IoT can help with inspections. For example, drones with cameras or sensors can inspect work to ensure quality and compliance with safety standards.

- *Elevated Efficiency:* IoT technology optimizes construction processes, reducing downtime and maximizing resource utilization. Real-time tracking and predictive maintenance are key examples of how IoT enhances efficiency.

- *Cost Reduction:* IoT technology helps in advanced budgeting and cost management. By monitoring and optimizing the use of resources, construction companies can save significantly on their budgets. Optimized resource use and predictive maintenance reduce operational costs.

5.7 CHALLENGES

The application of IoT in the real world is still limited by restraints, including limited battery life, network capacity, and maintenance cost. Although there can be different predictions regarding the potential job losses, implementing IoT technology will not replace humans in the construction industry. Adopting IoT in construction sector operations is neither a simple nor inexpensive procedure. Other challenges include the following [15,20]:

- *Data Privacy:* This could arise if IoT devices are not properly secured. Hackers increasingly target IoT devices because their security systems are not as robust as computers, phones, or other devices.

- *Legacy Systems:* Holding on to outdated technology slows down the construction industry exponentially. Obsolete hardware-software systems are expensive to maintain, lead to missed financial opportunities, reduce employee productivity, and make it harder to assess business efficiency.

- *Network Security:* Connecting a large number of low-memory and low-power devices to the networks means providing more nodes for hackers to hack/attack the network through these connected nodes (devices). These devices are connected to the network with wireless links for gathering, storing, and transferring data, making them more prone to security threats. It is quite challenging to protect the network and its layers from external threats and attacks because of the large number of connected devices, connectivity to the Internet, and lack of installation and updating of strong malware protection software. All these weaknesses make IoT networks vulnerable to security threats.

- *Lack of Knowledge:* Fear of the adoption of IoT and failure to produce quality output is the consequence of a lack of knowledge about the advantages of IoT. Compared with large-sized companies, small-sized companies are more reluctant to adopt advanced technologies because of a negligible awareness of the potential merits and challenges associated with advanced technologies. Overall, there is a lack of educational resources to embark on IoT knowledge.

- *Lack of Trained Labor:* A critical challenge for the construction industry is to adopt IoT without sufficient training. Adopting IoT means a reduction in labor because workers are not aware of and trained to handle IoT. Because the construction sector lacks competent IoT supporters, investment in training personnel is required to fill the skills gap.

- *Lack of Policies:* The lack of policies and implementation guidelines due to the heterogeneity of connected devices is significantly impeding IoT adoption in the construction industry. The lack of standardization of IoT architectures and slow progress in IoT research and development of Construction 4.0 makes organizations often adapt to their architectures and technologies, which impacts the market.

- *Resistance to Change:* The construction sector is incorporating new technology and communication strategies, but it should not be overlooked that this is still a somewhat slow process. Many builders and contractors are oblivious to the need for such devices and are unwilling to invest in them or change.

- *Power Management:* One of the major concerns impeding IoT adoption in the construction sector is energy and power control of intercommunicating devices. Most IoT devices are handheld with inbuilt battery. The limited battery life of IoT devices poses a constant challenge for IoT networks. Companies must constantly keep an eye on the battery status of the battery, i.e., when it needs to be recharged or replaced.

- *Interoperability:* The integration of IoT with the conventional Internet framework and network scaling is another challenge. The IoT infrastructure is an amalgamation of different devices and sensors. This leads to incompatibility between IoT devices and systems and communication and sharing service problems. This could make it impossible to integrate IoT devices with your existing project management or analytics software.

- *Expenditure:* It is a common fear among industrialists that adopting IoT infrastructure means great expense because extra funds will be required to acquire technology, implement it, and then train their employees. It is assumed to be very costly for small construction companies to support heterogeneous interfaces from diverse platforms. Power bills, spectrum license charges, and daily expenditure charges of an IoT node further add to the expense of adopting IoT technology.

- *Training Requirements:* This could add to the costs and deployment time if tradespeople who use the IoT systems are not familiar with them.

Figure 5.9 displays some of these concerns or challenges related to IoT adoption [17].

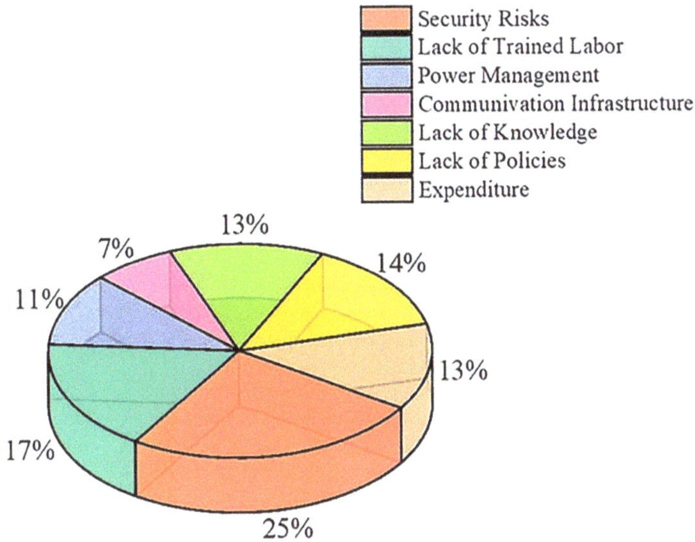

Figure 5.9 Some of the challenges related to IoT adoption [17].

5.8 CONCLUSION

The construction industry is in the research stage of the IoT adoption journey. In terms of IoT adoption, project managers and investors are still cautious to employ the technology; some are unsure how to utilize it, while others doubt its effectiveness. Things are changing as more construction companies are turning toward technologies like the IoT for help and many useful applications of IoT in construction have emerged. The construction industry, known for its inherent risks, has been embracing the IoT as a transformative tool to enhance safety measures on construction sites. IoT technology has proven to be a magnificent innovation in the construction industry by improving its working condition, reducing waste and increasing its efficiency. As IoT solutions for the construction sector become more omnipresent, they are having a huge impact on how the construction industry pivots.

The future of IoT in construction seems bright. Recent statistics show that IoT in construction will soon become a must-have technology. More information about Internet of things in the construction industry can be found in the books in [21-23] and the following related journal: *IEEE Internet of Things Journal*.

REFERENCE

[1] "Internet of things (IoT) in construction," May 2024,

https://projectvectors.com.au/posts/internet-of-things-iot-in-construction

[2] M. Suyogi, "How the Internet of things (IoT) can benefit the construction industry?" June 2021,

https://www.linkedin.com/pulse/how-internet-things-iot-can-benefit-construction-industry-suyogi/

[3] M. N. O. Sadiku, P. O. Adebo, and J. O. Sadiku, "Internet of things in the construction industry," *International Journal of Trend in Research and Development*, vol. 12, no. 1, January-February 2025, pp. 48-54.

[4] M. Townes, "The spread of TCP/IP: How the Internet became the Internet," *Millennium: Journal of International Studies*, vol. 41, no. 1, 2012, pp. 43 –64.

[5] "Internet of things (IoT),"

https://www.geeksforgeeks.org/internet-things-iot-2/

[6] M. N. O. Sadiku, and S.M. Musa and S. R. Nelatury, "Internet of things: An introduction," *International Journal of Engineering Research and Advanced Technology*, vol. 2, no.3, March 2016, pp. 39-43.

[7] P. Sadeghi et al., "Towards a reliable modulation and encoding scheme for Internet of things communications," *13th IEEE International Conference on Application of Information and Communication Technologies (AICT)*, October 2019,

https://www.researchgate.net/figure/Communications-Technologies-in-Internet-of-Things_fig1_335104959

[8] J. Rahm, "Internet of things in the manufacturing industry," April 2017,

https://blog.flexlink.com/internet-of-things-in-the-manufacturing-industry/

[9] M. N. O. Sadiku, Y. Wang, S. Cui, and S. M. Musa, "Industrial Internet of things," *International Journal of Advances in Scientific Research and Engineering*, vol. 3, no. 11, Dec. 2017, pp. 1-4.

[10] A. R. Sadeghi1, C. Wachsmann, and M. Waidner, "Security and privacy challenges in industrial Internet of things," *Proceedings of the 52nd Annual Design Automation Conference*, June 2015.

[11] "The industrial Internet of things (IIoT): The business guide to Industrial IoT,"

https://www.i-scoop.eu/internet-of-things-guide/industrial-internet-things-iiot-saving-costs-innovation/

[12] D. Schulz, "FDI and the industrial Internet of things," *Proceedings of IEEE 20th Conference on Emerging Technologies & Factory Automation*, 2015, pp. 1-8

[13] "IoT in construction market size, share, competitive landscape and trend analysis report, by application, by end user, by component: Global opportunity analysis and industry forecast, 2021-2031,"

https://www.alliedmarketresearch.com/iot-in-construction-market-A07565#:~:text=The%20Global%20IoT%20in%20Construction,14.6%25%20from%202022%20to%202031.

[14] D. Sinha, "Impact of IoT driven innovations in the construction industry," November 2019,

https://www.techaheadcorp.com/blog/iot-in-construction-industry/

[15] "The Internet of things in construction," March 2024, Unknown Source.

[16] "IoT in construction:Top benefits, use-cases & application,"

https://toolsense.io/equipment-management/iot-in-construction-top-benefits-use-cases-application/

[17] "9 Ways IoT makes construction more efficient," December 2024,

https://www.iotforall.com/9-ways-iot-makes-construction-more-efficient

[18] "IoT for enhanced safety in the construction industry," June 2023,

https://www.linkedin.com/pulse/iot-enhanced-safety-construction-industry/

[19] "Applying IoT in construction: Top 7 uses cases in 2024 and beyond," January 2024,

https://agilevision.io/blog/applying-iot-in-the-construction-industry-top-7-use-cases/

[20] K. Khurshid et al., "An in-depth survey demystifying the internet of things (IoT) in the construction industry: unfolding new dimensions," *Sustainability*, vol. 15, no. 2, 2023, p. 87

[21] M. N. O. Sadiku, *Internet of Things and Its Applications*. Moldova, Europe: Lambert Academic Publishing, 2024.

[22] W. Y. Long, *Smart City: Road of the modern urban construction in the context of the Internet of Things(Chinese Edition)*. Publishing House of Electronics Industry, 2000.

[23] A. Bhogayata and A. Sata, *Introduction to Internet of Things in Civil Engineering: Theory and Practice*. Springer, 2024.

CHAPTER 6
CLOUD COMPUTING IN CONSTRUCTION

"Cloud computing in construction is like having a digital blueprint accessible from anywhere, allowing teams to collaborate in real-time, streamline project management, and optimize resource allocation, paving the way for a more efficient and transparent building process."

—Anonymous

6.1 INTRODUCTION

Cloud computing (CC) is a model for enabling convenient, on-demand network access to a shared pool of configurable computing resources (e.g., networks, servers, storage, applications, and services) that can be rapidly provisioned. It just means storing and accessing information over the Internet rather than being tied to a computer's physical hard drive. Cloud computing technologies have revolutionized several industries including construction.

Construction is one of the oldest industries and it is also one of the most labor-intensive ones with more obligations. Yet, the construction industry has been notoriously slow to enter the digital era. Although the construction industry is well placed to leverage CC technologies for competitive and operational advantage, the diffusion of the technologies in the industry follows a steep curve. The construction industry is a very traditional sector. Figure 6.1 shows a typical construction site [1], while Figure 6.2 shows some construction workers [2]. Construction companies can streamline their operations and achieve greater efficiency through the use of cloud-based solutions, including project management tools, BIM platforms, and mobile applications.

Figure 6.1 A typical construction site [1].

Figure 6.2 Construction workers [2].

Cloud computing (CC) is a general term for anything that involves delivering hosted services over the Internet. It delivers computing services, including servers, storage, databases, software, networking and analytics, over the Internet ('the cloud'). It has emerged as a revolutionary technology that has transformed various sectors, including the construction industry. The uprise of cloud computing and smart technology is beginning to change everything about how construction is done, from beginning to end [3].

This chapter investigates the extent to which cloud computing has been used in the construction industry. It begins with explaining cloud computing basics. It describes cloud computing in construction and presents some of its applications. It highlights the benefits and challenges of cloud computing in the construction industry. It concludes with comments.

6.2 CLOUD COMPUTING BASICS

Cloud computing represents a newly emerging service-oriented computing technology. It is the provision of scalable computing resources as a service over the Internet. It allows manufacturers to use many forms of new p systems such as 3D printing, high performance computing (HPC), industrial Internet of things (110T), and industrial robots. It is transforming virtually every facet of modern manufacturing. It is innovating, reducing cost, and bolstering the competitiveness of American manufacturing [4].

The key characteristic of cloud computing is the virtualization of computing resources and services. Cloud computing is implemented in one of three major formats: software as a service (SAAS), platform as a service (PAAS), or infrastructure as a service (IAAS). These services are illustrated in Figure 6.3 [5] and explained as follows:

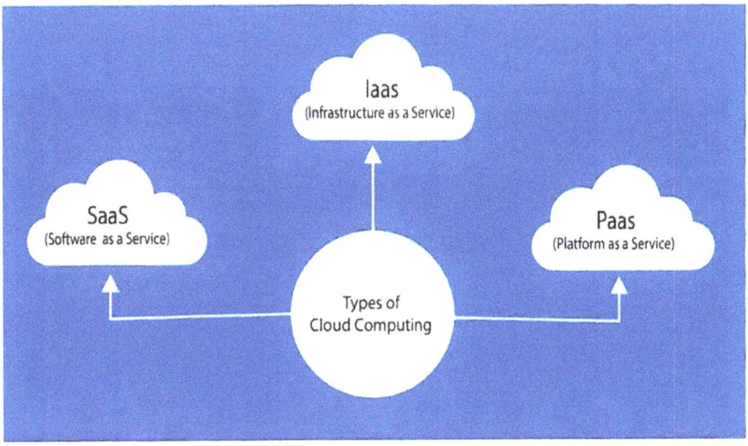

Figure 6.3 Three types of cloud computing services [5].

SaaS: This is a software delivery model in which software and associated data are hosted on the cloud. In this model, cloud service providers offer on-demand access to computing resources such as virtual machines and cloud storage. Nowadays oil & gas companies transition to cloud computing and implement SaaS solutions for operations.

PaaS allows the end-user to create a software solution using tools or libraries from the platform service provider. In this model, cloud service providers deliver computing platforms such as programming and execution.

In the IaaS model, cloud service providers can rent manufacturing equipment such as 3D printers.

Just like cloud computing, CM services can be categorized into three major deployment models (public, private, and hybrid clouds) [6]:

- Private cloud refers to a centralized management effort in which manufacturing services are shared within one company' or its subsidiaries. A private cloud is often used exclusively by one organization, possibly with multiple business units.
- Public cloud realizes the key concept of sharing services with the general public. Public clouds are commonly implemented

through data centers operated by providers such as Amazon, Google, IBM, and Microsoft.

- Hybrid cloud that spans multiple configurations. and is a composed of two or more clouds (private, community or public), offering the benefits of multiple deployment modes.

These models are shown in Figure 6.4 [7]. Cloud computing finds application in almost every field.

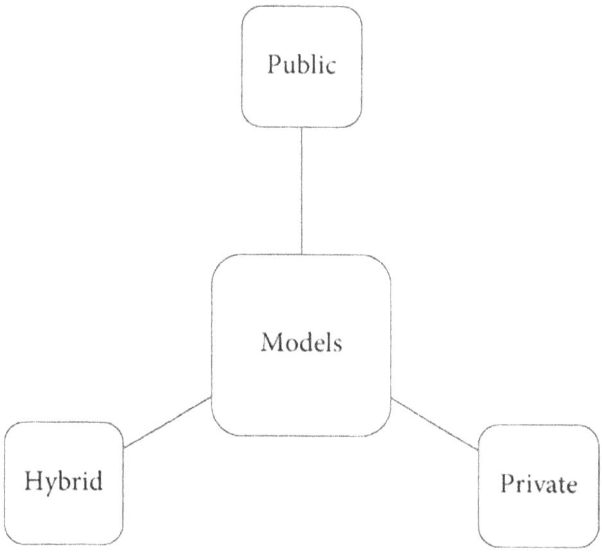

Figure 6.4 Cloud computing models [7].

6.3 CLOUD COMPUTING IN CONSTRUCTION

Cloud computing refers to the delivery of computing services, such as storage, processing power, and software applications, via the Internet. It allows you to access everything from your computer, tablet, or phone because it is all stored on the cloud (that is, on the Internet) and you can access your files from anywhere. The majority of construction work must be done at construction sites, the plant, the roads, and everywhere else in between. The construction industry is well-positioned to benefit from the CC technologies for operational and competitive advantages, but their widespread adoption follows a steep trend. As construction projects become increasingly complex, demanding seamless collaboration, efficient project management, and secure data storage, cloud computing offers a multitude of benefits and solutions. By integrating CC technologies, construction companies can enhance productivity, streamline processes, and achieve unprecedented levels of efficiency.

The use of cloud computing in the construction industry is a relatively recent development. Cloud computing in construction industry has become an active area of research evidenced by the soaring interest in the field. Cloud computing used in the construction industry has several models: private cloud, public cloud, and hybrid cloud. Public cloud is the most widely used cloud computing model. Examples of new business models for construction that could emerge include design-as-a-service, procurement–as-a-service, construction-as-a-service, facility management-as–a–service, demolition-as-a-service, retrofitting-as-a-service, waste management-as-a-service. Construction will thus be conceived as a utility and paid for per usage. Figure 6.5 shows cloud computing in construction [8].

Figure 6.5 Cloud computing in construction [8].

Cloud service providers create massive data centers containing hundreds of thousands of computers employed virtualization technology fortified with redundant power, networking, and connectivity. Popular cloud providers include Amazon, Google Inc., Microsoft Corp, Salesforce, Procore, CDC Software, APTEAN, and Oracle.

6.4 APPLICATIONS OF CLOUD COMPUTING IN CONSTRUCTION

Cloud computing can be applied to the construction industry in various aspects, including but not limited to architectural design, structural analysis, cost estimating, project planning and control, and procurement management. Cloud computing is used at different stages of the project life cycle, which include feasibility, design and construction stages. It is widely used in the construction industry for a variety of purposes, including project management, data storage, and collaboration. Figure 6.6 displays some applications of cloud computing in construction industry [9]. Common areas of applications of CC in construction include the following [9]:

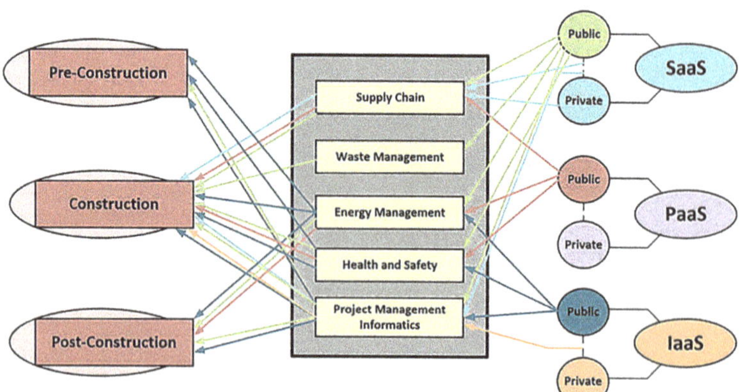

Figure 6.6 Some applications of cloud computing in construction industry [9].

- *Waste Management:* The construction industry can have a negative impact on the environment, with a staggering amount of materials and resources going to waste. Uncoordinated construction management is characterized with untimely feedback that results in wastage of resources. Wastage throughout the life cycle of a building must be minimized to reduce project cost and time over run. It has been observed that limited accessibility to existing construction information has resulted in resources wastage. Thus, there is a need for cloud computing to reduce transaction costs and enhance online collaboration tools.

- *Safe Construction:* Construction sites are usually dangerous due to a large number of workers, materials, equipment, and dynamic/unforeseen circumstances. The risk of working in the construction site is further increased by a lack of access to real time safety information to provide predictive, quantitative, and qualitative measures allowing the identification, correlation, and elimination of hazards and safety incidents. Cloud technologies are being employed to provide timely access to safety information and in turn resulting in improved safety performance of construction sites. Existing manual construction-safety monitoring is labor-intensive and error-prone. Thus a SaaS application is developed to detect unsafe conditions and analyze the trajectories of workers with respect to potential safety hazards on construction site.

- *Energy Management:* Buildings consume appreciable amount of energy during both construction and operation stages. Cloud computing has been employed to effectively manage energy in various stages of construction. The SaaS cloud has been proposed as an energy management system for sustainable decision support.

- *Supply Chain Management:* The existing uncoordinated traditional material supply in construction site usually resulted in supply gap leading to delay in projects. The public and private SaaS cloud platform has been employed for interoperability to bring together different stakeholders in the procurement process for efficient construction process.

- *Project Management Informatics:* The construction industry is characterized with communication and coordination problem culminating in low construction quality. Cloud computing technology for collaborative design will result in improved design, construction and project efficiency. The public SaaS cloud has been employed to formalize the transfer of knowledge among local construction companies to improve the construction project.

- *Collaboration:* In today's fast-paced world, construction projects have become more complex than ever before, requiring collaboration across different teams. Construction projects are executed by several project teams, with different business reporting models that are stored in different silos. Cloud computing enables construction teams to collaborate more effectively by enabling team members to access and share files and data in real time from anywhere with an Internet connection. The cloud makes it possible for everyone to be on the same page, no matter where they are. The word "team" has taken on a whole new meaning in the construction industry. Every team needs to obtain a high level of transparency and accountability for the property owner. Today's construction professional needs to be highly collaborative. Construction companies will be more collaborative, using technology to connect all project team members in one central location. By collaborating together on cloud platforms, construction stakeholders can store and receive construction data in real time.

- *Security:* Cloud computing providers have robust security measures in place, protecting data from loss or theft. Cloud security has matured and popular security measures in the cloud includes; encryption, use of up-to-date security software, cyber insurance security cover, security audit, and so on. Cloud service providers implement advanced security measures and best practices to ensure the safety of data stored in the cloud. This includes encryption, multi-factor authentication, and disaster recovery solutions. This way, companies avoid costly data breaches and other security issues. Cloud computing makes it easy to store fire safety plans, OSHA regulations and other important information in a central location so that everyone can access it at all times.

6.5 BENEFITS

The exploitation of cloud computing in construction project management has several advantages. The specific advantages of CC such as scalability, streamlining data resources, inclusive maintenance, etc. that make it an interesting concept for usage with the construction industry. CC helps to reduce costs, improve efficiency, collaboration and return on investment for developers and architects. Other benefits include [9,10]:

- *Back-office Functions:* These functions include billing, paying invoices, running financial reports, payroll, and planning logistics. They may have required physical presence on construction sites and are made possible using cloud computing. With remote access, construction sites are decentralized, while workers can work from various areas of the nation or even globally.

- *Sustainability:* Cloud computing adoption by construction companies is a great move towards a green society or a sustainable environment. Using cloud computing, construction companies can reduce their carbon footprint by eliminating the need for physical storage space and reducing paper usage. This can help them achieve their sustainability goals. This makes it so that both the customer and the contractor can feel good about their project's environmental impact without sacrificing efficiency or quality.

- *Improved Quality:* The quality of construction projects has come under scrutiny in recent years, with some high-profile failures leading to costly lawsuits. In the past, teams have had to rely on paper-based systems to track quality data. This reliance on paper tracking has led to errors and inconsistencies that employees have a difficult time correcting. With the cloud, all data is stored in a single, easily accessible location, making it easier to spot and fix any problems. Cloud computing can help improve the quality of your jobs by making it easier to track and compare data.

- *Accessibility:* The most useful feature of the cloud is that it is accessible from anywhere, anytime. Cloud computing enables remote access to data, making it easier for teams to work from different locations. It allows you to view job sites from anywhere. Using aerial photos or drone footage, cloud computing lets stakeholders view job sites from anywhere. This can help them better understand the project's progress, identify potential issues, and make informed decisions. For companies with various ongoing projects, the ability to access paperwork and information without being bound to a desk can save a lot of time and many trips back to the office.

- *Flexibility:* Cloud computing allows construction companies to access their data and applications from anywhere with an

Internet connection, making it easier to work remotely and helping construction companies with geographically dispersed teams or those needing to work on-site at different locations.

- *Cost Saving:* Cost is a significant barrier in the adoption of IT solutions by construction companies, because of its low-profit margin. Construction companies are seeking new ways to drive down infrastructure and operational costs. Cloud computing offers construction companies the opportunity to reduce both hardware and software expenses. By eliminating the need for extensive on-site infrastructure, significant capital investments can be saved. Cloud computing technologies have therefore provided opportunity to construction businesses especially SMEs to have access to high end computing infrastructure and applications which could cost a fortune to acquire. This will also undoubtedly translate to a reduction in the total cost of a project delivery, therefore giving construction companies a competitive advantage and operational edge.

- *Competitive Advantage:* A smaller construction company could find it difficult to compete with the giants of the industry, but such small companies could get a competitive edge with the help of cloud. Through cloud computing, even a small firm can achieve enterprise-level of recovery, management, access, and data storage at a fraction of the cost spent by bigger enterprises.

- *Scalability:* Cloud computing allows construction companies to scale their resources up or down as needed, making it easier to manage fluctuating workloads. Cloud computing enables a construction company to purchase IT resources as services dictated by the specific requirement at that particular period on a construction project. Cloud computing offers high-performance servers with powerful CPUs, GPUs, and super-fast SSD drives to construction industries at affordable prices.

- *Massive Storage:* Massive data generation characterizes construction projects, right from the design stage. The use of emerging technologies like artificial intelligence, IoT, Augmented reality, BIM generates continuously large data. For example, an aerial imagery of a site that will occupy points on a cloud storage, will take hundreds of GBs on a typical computer. Storing construction data on site requires physical access, whereas with cloud storage, data can be remotely stored and retrieved with no limitation to space and time. Hence the availability of cloud storage is an excellent opportunity for the construction industry.

- *Improved Project Management:* Cloud computing offers a centralized hub for storing and accessing project data, ensuring that all stakeholders can access the most current information. Cloud-based project management tools facilitate efficient scheduling and resource allocation, optimizing the utilization of labor, equipment, and materials. Real-time updates and notifications enable project managers to make timely adjustments, reducing downtime and maximizing productivity.

6.6 CHALLENGES

Although CC has brought about a revolution in the construction industry, it comes with its own set of challenges and limitations. There are also some adverse issues that the CC technology proponents need to address before the commercial world at large will be comfortable in its adoption. Some of these issues include trust, security, interoperability, and migration cost. No doubt, some construction applications characterized with fast processing and quick response time may not be able to rely on the distant and centralized cloud computing. Other challenges include [9,11]:

- *Resistance to Change:* Transitioning to cloud-based systems may require training and efforts in change management to familiarize employees with new tools and processes. It is important to address any resistance to change from stakeholders who are accustomed to traditional methods during the implementation phase.

- *Legacy Systems:* A legacy system is an older or outdated technology or software application that continues to be used because it still has value to the organization and functions as it was originally intended. Migrating data from legacy systems to the cloud and integrating existing systems with cloud-based solutions can be complex and time-consuming. Construction companies should carefully plan their data migration and integration strategies to minimize disruptions and ensure a seamless operation.

- *Interoperability:* Cloud vendors will have to adopt standards-based technologies in order to ensure true interoperability. Interoperability issues present themselves in different forms both at the system as well as information and data levels. Thus, the interoperability issue in a cloud computing model is more complicated in context. Cloud environments that are neutral to programming languages and operating systems are desirable. Interoperation among applications should be allowed inside a single cloud environment. This requires connectivity of the applications as well as semantics alignment of the data exchanged among the applications.

- *Threats:* There are threats to cloud-based computing technologies that may hinder the efficient adoption of cloud-based computing technologies in construction project delivery. Construction stakeholders find them significant to hindering the successful integration in the construction process. The main threats to the use of cloud-based computing technologies can be categorized into security threats and connectivity threat.

- *Latency:* Cloud adoption in construction may not guarantee acceptable transfer rate and response time required for some time sensitive construction applications. This could either be a software issue or network problem. Construction companies may avert delay by choosing service providers with closer data centres as fewer hops between the service provider and the customer. Overcoming latency and some barriers in cloud adoption will facilitate seamless connectivity that could evolve a real-time collaborative practice leading to collaborative computing. Figure 6.7 depicts categories of factors that cause construction delays [17].

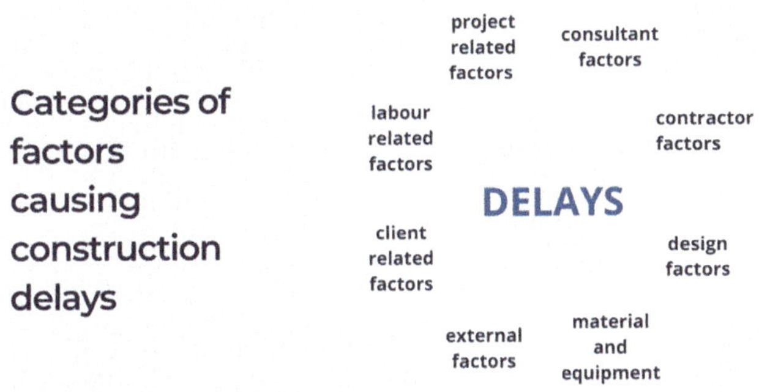

Figure 6.7 Categories of factors that cause construction delays [12].

- *Data Privacy:* There are concerns about data ownership and privacy in cloud computing, particularly when it comes to sensitive project data. The anticipated vulnerability in the adoption of cloud technology is increasing due to the increasingly fluidity of the security perimeter. In practice, business partners are usually unwilling to give their private and commercial information such as project cost to a third-party.

- *Data Availability:* It is not uncommon to experience down time with technologies utilizing cloud resources, as a technology cannot be perpetual. A cloud provider may have cause to shut down their resources unexpectedly. The building data thus becomes unavailable, what will become of the building data? To overcome this concern, the cloud providers are known to provide 99.999% availability as stated in the Service Level Agreement. This is to allow exchanging data between clouds providers to avoid the issue of lock-in if a provider became unavoidably unavailable.

- *Data Governance:* Since construction projects usually involve many professionals, there is the need to spell out the contractual relationship between stakeholders. This is critical since the data is being contributed, it may be regarded as being owned by all. The issue of actual owner of data may arise, since all concerned party has access to the data and are required to update data continuously. The relationship between the building team members is also another point to consider.

- *Poor Connectivity:* Access to cloud services is primarily over the Internet, hence, to maximize the benefit of cloud solution in the construction site, Internet connectivity must be available every time. The reliance on Internet connectivity for cloud-based services can pose a challenge, especially in remote construction sites that have limited access to reliable internet connections. To minimize disruptions caused by Internet outages, construction companies need to ensure they have alternative connectivity options or offline capabilities. Poor network connectivity is among the threats to cloud adoption by construction industry. Project sites at times might be an underdeveloped area or a rural area usually with low or no Internet connectivity.

- *Cost:* Accumulated cost for the use of cloud infrastructure over a long period could be daunting, depending on the type of deployment. Cloud SaaS pricing at inception was initially per unit time of consumption, further efforts by cloud researchers incorporated market value and brought about personalized pricing resulted in many pricings offers. The cost implication is not the same for the various cloud deployments types even for the same construction company. There could be substantial high cost of renting high-end resources such as GPUs for performing project analytics and machine learning tasks. Cost flexibility offered by cloud computing allows construction SMEs operating on tight budget to avoid significant upfront investment and operational cost.

- *Compatibility:* Cloud computing requires a certain level of technological sophistication, and not all construction companies may have the necessary infrastructure or skills to adopt it.

- *Data Security:* Security concerns are a common cause of hesitation about transferring data to the cloud. It can feel like you are uploading a lot of sensitive information such as intellectual property, plans, drawings, or financial data available for everyone to find online. Breaching a client's on-premise systems was generally easier than accessing their cloud-based data. You can ensure that no one can access your data unless you permit it by allowing access only to authorized users with unique login credentials.

- *Complexity:* Construction projects are getting increasingly complicated. As teams need to account for progressively complex planning and designing, the number of parties involved and stakeholders multiply from one project to the next. Construction companies are uniquely oriented to benefit from the cloud's ability to provide greater freedom and ease to access information anytime, anywhere.

6.7 CONCLUSION

Cloud computing has emerged as a transformative force in the construction industry, offering countless benefits such as cost savings, improved collaboration, and enhanced project management. It will result in emergence of new business models which will enable construction companies to do business in a new way. Almost as many concerns about abandoning traditional methods are in favor of the cloud. As the technology continues to evolve, we can expect to see further innovations in data management and collaboration. Implementing cloud tools could put you one step ahead of the competition.

In recent years, the interest in cloud solutions in construction has been booming. In spite of challenges such as limited Internet connectivity and resistance to change, the future of cloud computing holds a bright future for the construction industry. More information about cloud computing in construction can be found in the books in [13,14].

REFERENCE

[1] A. Frangoul, "Construction sites worldwide attempt to harness automated tech and other innovations," December 2020,

https://www.cnbc.com/2020/12/15/construction-sites-look-to-automated-tech-other-innovations-.html

[2] "Top reasons to consider a career in the construction industry," December 2022,

https://waynebrothers.com/construction-career-infographic/

[3] M. N. O. Sadiku, P. A. Adekunte, and J. O. Sadiku, "Cloud computing in the construction industry," *International Journal of Trend in Scientific Research and Development*, vol. 9, no. 1, January-February 2025, pp. 587-595.

[4] S. Ezell and B. Swanson, "How cloud computing enables modern manufacturing," June 2017,

https://itif.org/publications/2017/06/22/how-cloud-computing-enables-modernmanufacturing

[5] "Cloud computing applications in agriculture," https://www.eescorporation.com/cloud-computing-applications-in-agriculture/
[6] "Cloud manufacturing," *Wikipedia*, the free encyclopedia https://en.wikipedia.org/wiki/€loud_manufacturing

[7] J. S. Saini et al., "[Retracted] cloud computing: Legal issues and provision," *Security and Communication Networks*, vol. 2022,

https://www.hi dawi.com/jo rnals/scn/2()22/2288961/

[8] V. Prasad, "Cloud computing in construction: Enhancing collaboration and data management," May 2023,

https://www.linkedin.com/pulse/cloud-computing-construction-enhancing-collaboration-data-prasad/

[9] S. A. Bello et al., "Cloud computing in construction industry: Use cases, benefits and challenges," *Automation in Construction*,

vol. 122, February 2021.

[10] "How cloud computing is transforming the construction industry," April 2023,

https://multiproject.org/how-cloud-computing-is-transforming-the-construction-industry/#:~:text=Cloud%20computing%20enables%20construction%20teams,improve%20communication-%20and%20enhance%20teamwork.

[11] A. Afolabi et al., "The economics of cloud-based computing technologies in construction project delivery," *International Journal of Civil Engineering and Technology*, vol. 8, no. 12, December 2017, pp. 233–242.

[12] "5 Benefits of cloud computing for the construction industry,"

https://gocodes.com/construction-cloud-computing/?srsltid=AfmBOorcVCwmX9O-C0xyqYOrWpkYwYwZo8iyEH1RyVkXp6RPC3xuIqPw

[13] M. N. O. Sadiku, *Cloud Computing and Its Applications.* Moldova, Europe: Lambert Academic Publishing, 2024.

[14] B. Furht and A. Escalante, *Handbook of Cloud Computing.* New York: Springer Science Business Media, 2010.

CHAPTER 7
BLOCKCHAIN IN CONSTRUCTION

"The whole point of using a blockchain is to let people — in particular, people who don't trust one another — share valuable data in a secure, tamperproof way."

—MIT Technology Review

7.1 INTRODUCTION

Blockchain is a system that tracks transactions across a peer-to-peer network. In some cases, these transactions are made via either cryptocurrency or Bitcoin. Blockchain technology (BCT) is a revolutionary tool used to record transactions and link them together to form a "chain" that's known as a distributed ledger. It was introduced a decade ago as the underlying technology behind Bitcoin. BCT application has been extended since its first appearance in 2008 by Satoshi Nakamoto, from financial services to agricultural supply chains. Blockchain has the potential to revolutionize many different industries, including construction. Blockchain in construction, which was initially developed to support cryptocurrencies, has transcended its intended use and found use in a variety of industries, offering increased transparency, improved security, and simplified procedures. Figure 7.1 shows the symbol of blockchain [1].

Figure 7.1 The symbol of blockchain [1].

The construction industry happens to be one of the most critical industries in the world. We rely on construction workers to build everything from roads and bridges to hospitals and schools. Yet, the construction industry has traditionally been very slow to adopt new technologies. Integrating recent technologies such as the Internet of things (IoT), artificial intelligence, and blockchain creates the potential for overcoming many limitations within the construction industry. Considering the challenges faced in the construction industry, including schedule delays, cost overruns, and quality control issues, blockchain can potentially revolutionize project planning, design, execution, and long-term maintenance, thereby improving transparency, efficiency, and collaboration in the construction process. Figure 7.2 shows a typical construction site [2], while Figure 7.3 shows some construction workers [2].

Figure 7.2 A typical construction site [2].

Figure 7.3 Some construction workers [2].

Blockchain is a distributed database that allows for secure, transparent, and tamper-proof transactions in its simplest form. It is a system that tracks transactions across a peer-to-peer network. Its main defining feature is decentralization. It is poised to transform the entire foundation of the construction sector, which is on the verge of a technological revolution. It can solve the problems rooted in payment systems, collaboration, documents in the construction industry [3].

This chapter explores how blockchain technology could potentially change the construction industry for the better. It begins with explaining what blockchain is all about. It describes blockchain in construction and provides some of its applications. It highlights the benefits and challenges of blockchain in the construction industry. It concludes with comments.

7.2 WHAT IS BLOCKCHAIN?

Blockchain, a type of distributed digital ledger technology (DLT), is a relatively new and exciting way of recording transactions in the digital age. It is a decentralized and distributed digital ledger technology that securely records and verifies transactions across multiple computers or nodes in a network. Basically, it is a chain of blocks in which each block contains a list of transactions. The blockchain technology was created as the foundational basis for Bitcoin – a digital currency in which secure peer-to-peer transactions occur over the Internet. It is expected that the spending on blockchain solutions worldwide would grow from 4.5 billion USD (2020) to an estimated value of 19 billion USD by 2024 [4].

Originally developed as the accounting method for the virtual currency Bitcoin, Blockchains are appearing in a variety of commercial applications today. Blockchain technology is a type of distributed digital ledger that uses encryption to make entries permanent and tamper-proof and can be programmed to record financial transactions. It is used for secure transfer of money, assets, and information via a computer network such as the Internet without requiring a third-party intermediary. It is now being adopted across financial and non-financial sectors. As a catalyst for change, the Blockchain technology is going to change the business world and financial matters in major ways.

The first Blockchain was conceived in 2008 by an anonymous person or group known as Satoshi Nakamoto, who published a white paper introducing the concept of a peer-to-peer electronic cash system he called Bitcoin [5,6]. Bitcoin and Ethereum are the first two mainstream blockchains. Other modern blockchains include Namecoin, Peercoin, Ether, and Litecoin. Figure 7.4 shows different components of blockchain [7].

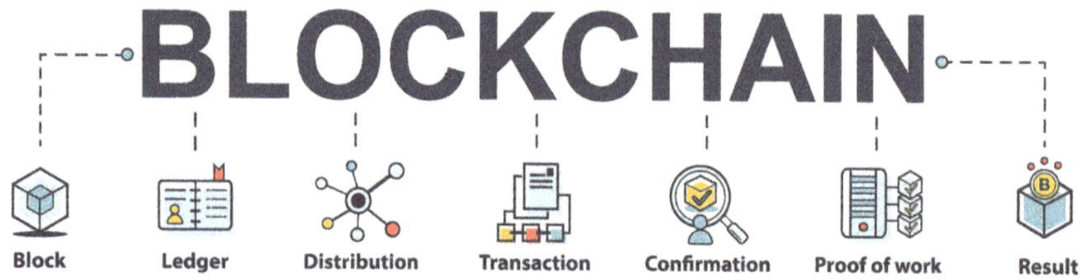

Figure 7.4 Different components of blockchain [7].

Blockchain combines existing technologies such as distributed digital ledgers, encryption, immutable records management, asset tokenization and decentralized governance to capture and record information that participants in a network need to interact and transact. As illustrated in Figure 7.5, a complete blockchain incorporates all the following five elements [8]:

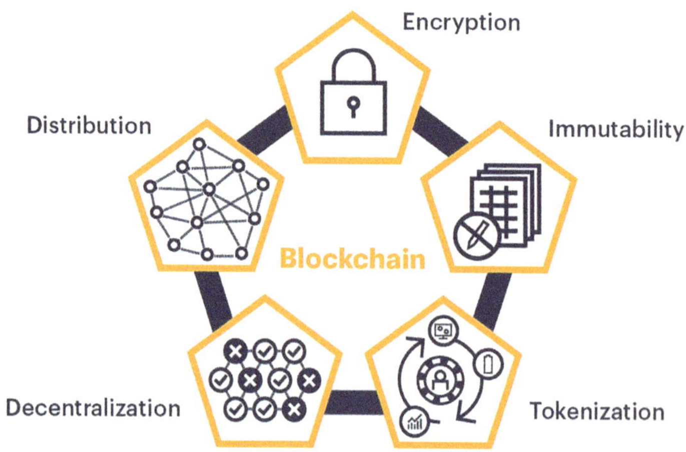

Figure 7.5 Five key elements of Blockchain [8].

- *Distribution:* Digital assets are distributed, not copied or transferred. A protocol establishes a set of rules in the form of distributed mathematical computations that ensures the integrity of the data exchanged among a large number of computing devises without going though a trusted third party. A centralized architecture presents several issues including a single point of failure and problems of scalability.

- *Encryption:* BC uses technologies such as public and private keys to record data securely and semi-anonymously. Completed transactions are cryptographically signed, time-stamped, and sequentially added to the ledger.
- *Immutability:* The blockchain was designed so these transactions are immutable, i.e. they cannot be deleted. No entity can modify the transaction records. Thus, Blockchains are secure and meddle-free by design. Data can be distributed, but not copied.
- *Tokenization:* Value is exchanged in the form of tokens, which can represent a wide variety of asset types, including monetary assets, units of data or user identities.
- *Decentralization:* No single entity controls a majority of the nodes or dictates the rules. A consensus mechanism verifies and approves transactions, eliminating the need for a central intermediary to govern the network.

Bitcoin and its underlying blockchain technology increasingly impact all facets of society. Bitcoin's status as digital gold is merely the tip of this technology. Figure 7.6 shows Bitcoin [9], while Figure 7.7 shows how blockchain works [10]. Although blockchain technology will for all time be associated with Bitcoin due to their common genesis, it has broader applications. Cryptocurrency will increasingly become a factor in family law issues as well.

Figure 7.6 Bitcoin [9].

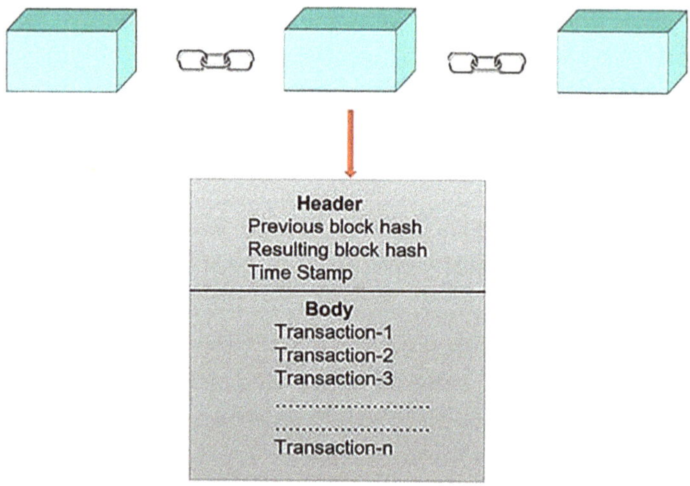

Figure 7.7 How blockchain works [10].

A blockchain is a tamper-proof, distributed database that stores blocks of information for cryptographically bound transactions via peer-to-peer networks. At the heart of blockchain's functionality is cryptographic hashing. Each block in a blockchain contains a cryptographic hash of the previous block, creating an immutable chain of blocks. If anyone attempts to tamper with the data in a block, it would alter the block's hash. This would disrupt the entire chain, making it virtually impossible to manipulate. The security feature ensures data integrity and prevents unauthorized changes

[11].

In a nutshell, blockchain technology involves three basic concepts [12]: (1) It is a system for recording a series of data items (such as transactions between parties); (2) It uses cryptography to make it difficult to tamper with past entries; (3) It has an agreed process for storing copies of the ledger and adding new entries (also called a consensus protocol).

Blockchain is a novel decentralized infrastructure and distributed computing paradigm that uses a chained data structure for verification, storage, and distributed consensus algorithms to generate and update data. Decentralization is a key feature of blockchain technology, which refers to the distribution of power and decision-making across a network of nodes or participants rather than being controlled by a central authority or system. It provides robustness while eliminating many-to-one traffic flows to avoid delays and single points of failure. Figure 7.8 shows the decentralized property of blockchain [10].

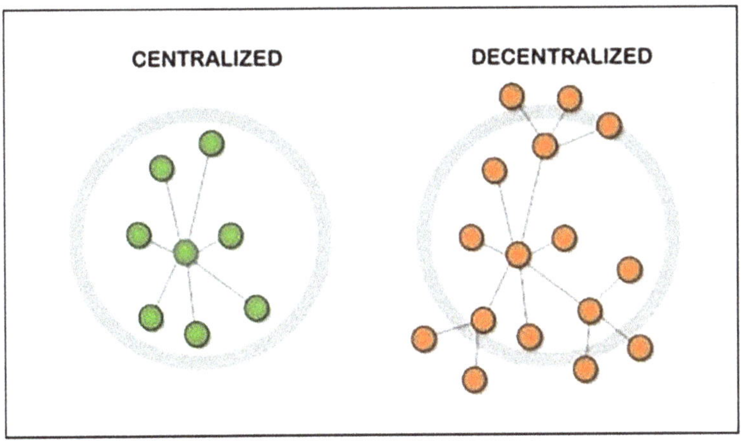

Figure 7.8 The decentralized property of blockchain [10].

The advantages of decentralized property of blockchain network include the following [10]:

- The decentralized property of blockchain makes it less prone to failure and more expensive for hackers to attack the network.

- There is no third-party involvement; therefore, there is no added risk.
- Every change made in the network is traceable and concrete.
- Users maintain full autonomy of their properties and are not dependent on third parties to maintain and manage their assets.
- It provides enhanced security.

7.3 BLOCKCHAIN IN CONSTRUCTION

Blockchain in construction means applying blockchain technology in the construction process. The use of blockchain in construction holds enormous promise for revolutionizing the way construction projects are planned, carried out, and controlled. A blockchain in construction is fundamentally a distributed, unchangeable digital ledger that securely and openly records transactions and other data. It has the ability to transform supply chain management, automate contract execution, guarantee document authenticity, and promote effective stakeholder engagement because of its decentralized and transparent nature. You can use blockchain in the construction industry to modernize project management, track supply chains, and handle financial transactions. Figure 7.9 shows blockchain in construction [13].

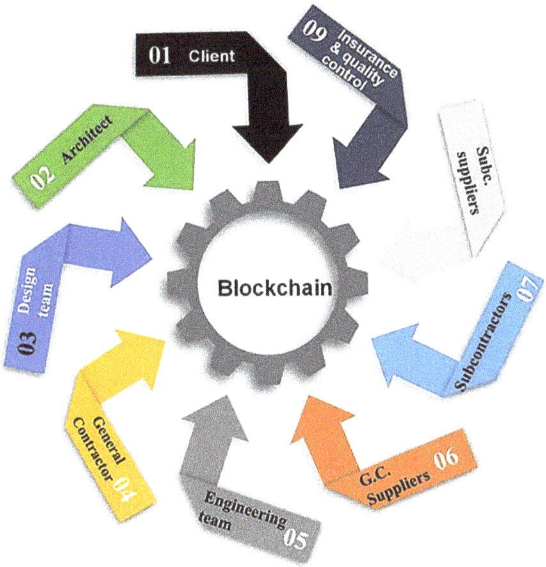

Figure 7.9 Blockchain in construction [13].

Although blockchain as a technology has been around for some time, the very idea of integrating it into construction is new. Construction is one of several industries that blockchain technology has the potential to disrupt. With its built-in attributes of transparency, security, and decentralization, blockchain can help the construction industry overcome significant obstacles, increase productivity, and simplify procedures.

7.4 APPLICATIONS OF BLOCKCHAIN IN CONSTRUCTION

Blockchain technology (BCT) is being used in the building industry, from quality control and supply chain management to document verification and project collaboration. BCT has vast potential applications in the construction industry, such as supply chain management, project bidding, contract management, management of certifications and permits, transparency and traceability, supplier management, quality control, certification, smart contracts and automation, streamlined payment processes, immutable contract execution, dispute resolution, document management and authentication, secure sharing and storage, digital signatures and timestamps, anti-fraud measures, project management, etc. Common areas of application include the following [2,14,15]:

- *Construction Payments:* Delayed payments are a frequent problem in the construction industry. This can create a lot of financial strain for construction companies, as they often have to borrow money to cover materials and labor costs. By facilitating direct and secure peer-to-peer payments, blockchain-based payment systems can simplify financial transactions in the construction industry. Blockchain technology can be used to streamline construction payments. With blockchain technology, construction companies could receive payments almost instantly after completing a job. Blockchain improves the speed and efficiency of payment operations by doing away with middlemen and cutting transaction costs.

- *Construction Contracts:* Another way that blockchain technology could be used in construction is through smart contracts. Smart contracts are largely hailed as the future of the construction industry. A smart contract is a digital contract that is stored on the blockchain. Smart contracts are autonomous contracts with set terms. They are immutable, meaning they cannot be altered or deleted. With a smart contract, the terms of the contract would be clearly defined and could not be changed. In construction projects, smart contracts based on BCT can enhance payment transparency, protect transaction histories, and provide secure access to information. Smart contracts boost efficiency by doing away with the need for middlemen and manual processing, which lowers costs, delays, and costs associated with project execution. Figure 7.10 shows smart contracts [16].

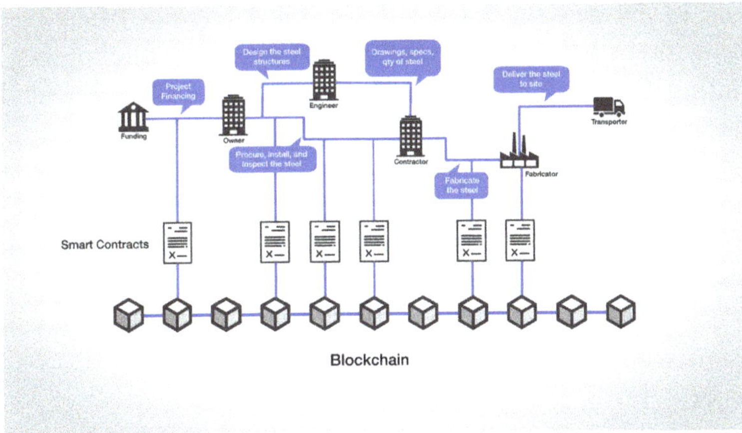

Figure 7.10 Smart contracts [16].

- *Project Management:* Blockchain technology can be used to improve project management in construction. Construction projects are often complex and involve many different parties. With blockchain technology, all of the parties involved in a project would be able to view the digital ledger and see the progress made. This would help to improve communication and coordination between the different parties.

- *Procurement Process:* Blockchain technology can also be used to streamline the construction procurement process. In construction, procurements are often slow and cumbersome because many different stakeholders are involved in the process, including architects, engineers, contractors, and suppliers. With blockchain technology, all stakeholders would be able to view the digital ledger and see the procurement status.

- *Disputes:* Disputes in organizations with multiple stakeholders are inevitable. In a competitive environment such as construction, there are potential conflicts between participants whose views and benefits are different. Claims, which are inevitable in construction projects, turn into disputes unless managed. Disputes are rooted in three general fields: payment, document, and collaboration. BCT can manage various systems challenges by offering decentralization, security, and transparency.

- *Supply Chain Management:* The construction industry is

known to have complex, lengthy, and network-based supply chains with a vast range of internal and external stakeholders. The prevailing practices in construction supply chain management are characterized by significant challenges, including fragmentation, a lack of trust among stakeholders, and restricted information sharing. By offering a transparent and secure platform for tracking the movement of products and resources, blockchain can improve supply chain management in the construction industry. The blockchain can be used to record every transaction and movement of products, allowing participants to track the origin, location, and status of items in real-time.

- *Transparency and Traceability:* Blockchain is a distributed ledger that will retain a permanent, tamperproof transaction record, which is one of its main advantages for real estate. Should the transaction ledger be shared across the entire industry network, it will promote transparency and help minimize mistrust among industry players, speeding up the whole transaction process. Construction professionals gravitate towards blockchain technology because of the benefits of transparency. Transparency may be regarded as a blockchain that orders and sends transactions to all members in an apparent manner. Throughout the construction process, transparency and traceability are made possible via distributed ledger technology on the blockchain. As a result of the blockchain's ability to record every transaction, alteration, and update and make them accessible to authorized users, there is less room for fraud and manipulation.

- *Quality Control:* By offering a visible and unchangeable record of inspections, testing, and certifications, blockchain technology can improve quality control and certification procedures in the construction industry. The blockchain can securely record all pertinent data, including test results, compliance documentation, and the materials utilized. This guarantees that quality requirements are met. Quality control software standardizes quality control processes and ensures compliance with regulations and standards.

- *Data Immutability:* Data recorded in a blockchain are considered unchangeable, which is referred to as the immutability of the blockchain. The immutability of BCT can provide significant benefits to the construction industry by improving accountability, supply chain management, and regulatory compliance. The chances of fraud are reduced due to the blockchain's immutability. With the immutability of blockchain, data integrity, which plays an important role, can also be easily achieved.

7.5 BENEFITS

Since BCT platforms are decentralized, reducing unnecessary bureaucracy will improve open communications, and BCT application can prevent disputes rooted in communication issues, inaccurate documentation, and payment problems. The construction sector can overcome long-standing obstacles, boost production, and complete projects more successfully in the digital era by embracing blockchain. Other benefits include [2]:

- *Security:* This is an important aspect of blockchain technology. It is very challenging for hackers to tamper with the data kept on the blockchain because of the decentralized architecture of the network and the usage of cryptographic algorithms. The network is very resistant to fraud and criminal activity since consensus procedures are used to verify that everyone using it agrees on the authenticity of transactions.

- *Secure Storage:* Traditional methods of storing information involved a central repository from which data could be accessed from anywhere. This method is always prone to security attacks, that could alter or tamper with the transaction statements from the central repository. For critical construction data, blockchain offers a safe and decentralized storage solution. Blockchain disperses data among numerous nodes, making it highly resistant to illegal access or manipulation. The system for safe data transfer and storage improves privacy and security. Unlike traditional storage methods, which provide a central repository (particularly subject to security attacks) from which data can be accessed from anywhere, the blockchain is based on the creation of a distributed virtual distributed virtual register (tamper-proof), in which data is shared, stored and verified by the who manage the register in chronological order and without the need for an intermediary.

• *Anti-Fraud Measures:* The transparency and immutability of blockchain serve as effective barriers against fraud in the building sector. The blockchain keeps track of every transaction and modification, making it more difficult for bad actors to modify data or carry out fraudulent activities.

• *Project Tracking:* Project milestones, tasks, and resources may be accurately tracked because of the blockchain's transparency and audibility. Better project control, on-time delivery, and increased overall performance are all results of this improved project tracking.

• *Transparency and Trust:* The capacity of blockchain in construction to promote transparency and trust among all stakeholders is one of the most important advantages of this technology in the building industry. Blockchain in construction generates an immutable and transparent ledger of transactions in a typically fragmented business where information transmission is frequently time-consuming.

• *Reduced Fraud:* Blockchain technology for construction minimizes the fraud risk and transparency issues by introducing a decentralized, tamper-proof ledger system. In other words, the tamper-proof and secure nature of blockchain considerably lowers the chance of fraud and disagreements. Due to the cryptographical links between records kept on the blockchain in construction, it is very difficult to alter or corrupt data without being noticed. This quality is especially important in the construction industry, where disagreements frequently develop around project milestones, payments, and quality control.

• *Improved Efficiency:* The paperwork-heavy procedures common in the construction sector are known to cause delays, mistakes, and inefficiencies. The automated and digital method of blockchain in construction streamlines the paperwork procedure. This shortens project times, decreases errors, and reduces administrative burden.

• *Modernized Contracts:* Blockchain changes the way of contract management in construction. It uses smart contracts that automatically carry out agreed-upon terms, cutting out the middleman. Self-executing contracts with established rules and circumstances are known as smart contracts, another essential component of blockchain in construction. This automation makes sure that payments and deliveries only happen when certain conditions are met. It makes contracts more secure and efficient.

- *Increased Accountability:* Blockchain introduces a new level of responsibility in construction. Everyone involved gets access to the same unchangeable data, ensuring each party is held accountable for their actions. It leads to a more honest and responsible way of implementing construction projects.

Some of these benefits are illustrated in Figure 7.11 [10].

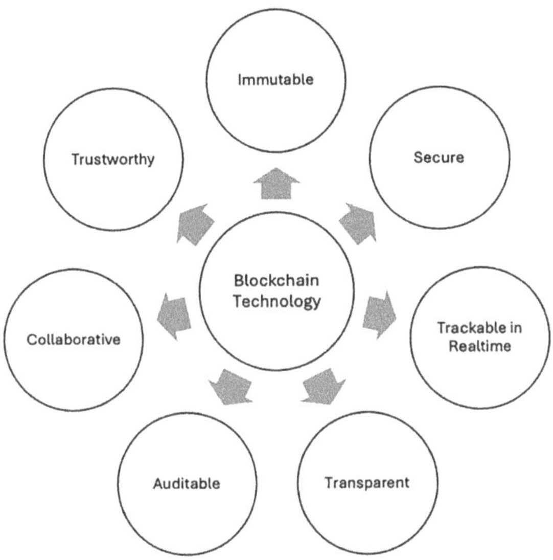

Figure 7.11 Some of the benefits of blockchain in construction [10].

7.6 CHALLENGES

Blockchain adoption in the construction industry is not without its challenges. Productivity within the construction industry is adversely affected by traditional contracting arrangements, with frequent suspicion and mistrust among collaborative partners and stakeholders. To achieve successful adoption of BCT in construction, a number of obstacles and factors must be taken into account. Obstacles that must be overcome include technological infrastructure, data privacy and security issues, and the requirement for industry-wide collaboration. Trust and transparency, cost overruns, complex operational management, fragmentation, payment delays, and on-time project completion are some of the other challenges facing the industry. Some of these challenges are explained as follows [2]:

- *Resistance to Change:* The construction industry has traditionally been very reluctant to change and rather slow in adopting new digitization technologies. It can be difficult to persuade stakeholders to invest in this technology, particularly if they are used to conventional approaches. It is essential to show the long-term advantages and return on investment in order to get over change resistance.

- *Cost:* In order to implement blockchain, there is a variety of complex systems that will likely need to be created. This comes at a cost, both for the systems and to hire the people necessary to create and implement them. Implementing blockchain technology can involve a significant upfront cost. This comprises the cost of data migration, development, integration, and training. The costs for new hardware, software, and training can add up quickly. Thus, it becomes hard for smaller companies to afford these upgrades.

- *Infrastructure:* To use blockchain, you need a strong tech setup, which many construction companies lack. Blockchain implementation in the building industry needs a strong technological foundation that can support the dispersed nature of the technology. The construction businesses must make sure they have the requisite hardware, network capabilities, and technical know-how to manage the demands of blockchain implementation.

- *Data Privacy and Security:* In a world where massive data is generated daily, data privacy has become a major concern. Blockchain goes a long way in addressing this issue. Given that the construction business deals with sensitive information and intellectual property, data privacy and security are major issues. Sensitive data is used in construction projects, such as confidential financial information and private designs. Blockchain technology enhances security through encryption and decentralization, but businesses must still take into account the privacy implications and make sure they are in compliance with all applicable laws. The cryptographic encryption used by blockchains improves data security by ensuring that private data is shielded from unauthorized access.

- *Collaboration:* Architects, engineers, contractors, and clients may communicate securely using blockchain-based collaboration tools. Blockchain improves cooperation, minimizes communication gaps, and promotes effective decision-making by offering a shared workspace for information exchange. Collaboration is necessary for the successful deployment of blockchain in the construction industry.

- *Standardization:* For best practices to be established across the sector and to encourage wider adoption, collaboration, and standardization efforts are essential. Standardization promotes widespread usage of blockchain technology in the construction industry and facilitates frictionless data interchange while increasing transparency. To create industry-wide standards, protocols, and governance models, coordination is required between construction businesses, technology providers, and regulatory organizations.

- *Legacy Systems:* Integration with existing systems and procedures is one of the main obstacles to using blockchain in construction. It can be challenging to modify blockchain in construction so that it can live with legacy systems while maintaining data compatibility and consistency. The creation of interoperability standards, careful planning, and customization are all necessary for seamless integration.

- *Scalability:* As you add more to a blockchain, it can slow down, becoming less efficient, especially for big projects where speed is key. Blockchain networks may experience scalability problems as more transactions are processed. Making sure the blockchain in construction can manage a lot of transactions at once is essential for the construction industry, which has many participants and generates a lot of data.

- *Sustainability:* Environmental concerns have also been raised by the energy consumption of blockchain networks, particularly proof-of-work-based systems. Sustainable blockchain adoption necessitates actions to address scalability and reduce energy consumption.

- *Legislation:* The idea of using blockchain is one thing, while the very process of integrating it into various industries is a different thing altogether. Lawmakers are yet to introduce legislation to make what happens on the blockchain legally binding. A lack of such legislation will hinder blockchain adoption across various industries. Legislation is required to create proper functioning of smart contracts and to govern how disputes relating to such contracts will be resolved.

- *Regulation:* The decentralized nature of blockchain puts the established legal and regulatory systems to the test. An intricate web of laws, agreements, and compliance requirements governs the construction business. It might be challenging to implement blockchain systems that are compatible with current regulatory frameworks and guarantee security and transparency.

- *Enhanced Transparency:* Blockchain brings unmatched clarity to construction projects. It records every action, from buying materials to signing contracts, on a secure, unchangeable ledger. This clarity cuts down the risk of fraud and poor management. As a result, stakeholders investing in these projects can be confident that their money is being used properly.

Figure 7.12 depicts some of the challenges and limitations of blockchain in construction [17].

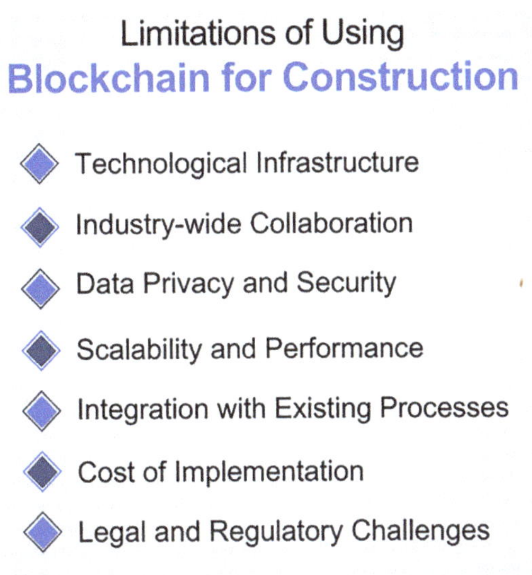

Figure 7.12 Some of the challenges and limitations of blockchain in construction [17].

7.7 CONCLUSION

Blockchain technology is among the most disruptive forces of the past decade. Its adoption is a trend that will help the construction industry become more efficient overall. The development of blockchain technology offers the construction sector a hitherto unheard-of possibility. Construction businesses may improve supply chain efficiency, cut down on delays, and minimize risks by embracing the transparency and traceability offered by blockchain technology.

In recent years, blockchain technology has attracted momentous attention in the construction industry. BCT has a bright future in the construction sector and has the ability to significantly impact the sector. It is on track to make a big impact on the construction industry. It promises to make things more efficient, transparent, and cooperative. More information on the integration of blockchain technology into the construction industry is available from the books in [18-20] and in the following related journals:

- *IEEE Blockchain*

- *Construction Magazine*
- *Buildings*
- *Journal of Building Engineering*
- *Journal of Construction Engineering and Management*

REFERENCE

[1] "Blockchain meets the oil & gas industry," February 2018,

https://executiveacademy.at/en/news/detail/blockchain-meets-the-oil-gas-industry/

[2] F. Farhadi, "Blockchain in construction: Ultimate guide for 2024," July 2023,

https://neuroject.com/blockchain-in-construction/

[3] M. N. O. Sadiku, P. A. Adekunte, and J. O. Sadiku, "Blockchain in the construction industry," *International Journal of Trend in Scientific Research and Development*, vol. 9, no. 1, January-February 2025, pp. 568-579.

[4] C. M. M. Kotteti and M. N. O. Sadiku, "Blockchain technology," *International Journal of Trend in Research and Development*, vol. 10, no. 3, May-June 2023, pp. 274-276.

[5] "Blockchain," *Wikipedia*, the free encyclopedia

https://en.wikipedia.org/wiki/Blockchain

[6] S. Nakamoto, "Bitcoin: A peer-to-peer electronic cash system,"

https://bitcoin.org/bitcoin.pdf

[7] "The beginning of a new era in technology: Blockchain traceability,"

https://www.visiott.com/blog/blockchain-traceability/#:~:text=The%20Beginning%20of%20a%20New,money%20without%20a%20central%20bank.

[8] "The CIO's guide to blockchain,"

https://www.gartner.com/smarterwithgartner/the-cios-guide-to-blockchain#:~:text=True%20blockchain%20has%20five%20elements,%2C%20immutability%2C%20tokenization%20and%20decentralization.

[9] "Blockchain and space exploration: Is decentralized data the future of space missions?" October 2024,

https://medium.com/coinmonks/is-decentralized-data-the-future-of-space-missions-646173d1aeec

[10] B. G. Celik, Y. S. Abraham, and M. Attaran, "Unlocking blockchain in construction: a systematic review of applications and barriers," *Buildings*, vol. 14, no. 6, 2024.

[11] D. Singh, "Blockchain in pharmaceutical supply chain: The next big frontier," October 2023,

https://www.debutinfotech.com/blog/blockchain-in-pharmaceutical-supply-chain-the-next-big-frontier

[12] D. Michels, "Technology blockchain and telecoms,"

https://www.iicom.org/wp-content/uploads/22-26-blockchain.pdf

[13] Z. Dakhli, Z. Lafhaj, and A. Mossman, "The potential of blockchain in building construction," *Buildings*, vol. 9, no. 4, 2019.

[14] "Will blockchain change the construction industry?" Unknown Source.

[15] W. Sun et al., "Critical success factors for implementing blockchain technology in construction," *Automation in Construction*, vol. 156, December 2023.

[16] K. A. Anand, " What is blockchain technology in construction?" October 2023,

https://www.getpowerplay.in/resources/blogs/what-is-blockchain-technology-in-construction/

[17] "Blockchain for construction: Enhancing efficiency & transparency," February 2024,

https://blog.webisoft.com/blockchain-for-construction/

[18] M. N. O. Sadiku, *Blockchain Technology and Its Applications*. Moldova, Europe: Lambert Academic Publishing, 2023.

[19] T. Dounas and D. Lombardi (eds.), *Blockchain for*

Construction. Springer, 2022

[20] F. Elghaish et al., *Blockchain of Things and Deep Learning Applications in Construction: Digital Construction Transformation.* Springer, 2023.

CHAPTER 8
3D PRINTING IN CONSTRUCTION

"3D printing has the potential to revolutionize the construction industry, enabling us to build homes faster, more affordably, and with less waste."

—Behrokh Khoshnevis

8.1 INTRODUCTION

A 3D printer is regarded as a disruptive technology that will change manufacturing. It has been used for decades in the automotive and aerospace industries. The 3D printer is also used by hobbyists, small businesses, creatives, manufacturers, architects, and most importantly contractors to instantly create a variety of products.

3D printing has made people awe-filled with its potential impact on the manufacturing industry. The construction industry is one of the most exciting applications of 3D printing technology because structures can be built using a fraction of the time, labor, and material compared to traditional structures. The construction industry is changing due to 3D printing, making it cheaper and faster to construct buildings and other structures by automating building processes. 3D printing can be a more cost-effective, convenient, and environmentally sound way to construct new buildings. It can solve many drawbacks of traditional construction. Some companies have used 3D printing to create offices, houses, and other structures. The use of 3D printing in construction is set to explode. In April 2021, the first prototype 3D printed house made out of clay, Tecla, was completed. Such buildings could be cheap, well-insulated, stable and weatherproof, climate-adaptable, and customizable [1].

3D printing, also known as additive manufacturing, is a process of creating a physical object from a digital model. Some companies are using 3D printers to revolutionize construction. There are many advantages to using 3D printing technologies in construction. With 3D printing, the argument for the pros and cons comes down to money and time, as well as the environmental benefits. The use of 3D construction printing has been growing rapidly in recent years [2].

This chapter examines the use of 3D printing technology in the construction industry. It begins with explaining what 3D printing is all about. It presents 3D printing in construction. It provides several applications of 3D printing in the construction industry. It highlights the benefits of and challenges of 3D printing in construction. It concludes with some comments.

8.2 WHAT IS 3D PRINTING?

3D printing (also known as additive manufacturing (AM) or rapid prototyping (RP)) was invented in the early 1980s by Charles Hull, who is regarded as the father of 3D printing. Since then it has been used in manufacturing, automotive, electronics, aviation, aerospace, aeronautics, engineering, architecture, pharmaceutics, consumer products, education, entertainment, medicine, space missions, the military, chemical industry, maritime industry, printing industry, and jewelry industry [3]

A 3D printer works by "printing" objects. Instead of using ink, it uses more substantive materials–plastics, metal, rubber, and the like. It scans an object–or takes an existing scan of an object–and slices it into layers, which can then convert into a physical object. Layer by layer, the 3D printer can replicate images created in CAD programs. In other words, 3D printing instructs a computer to apply layer upon layer of a specific material (such as plastic or metal) until the final product is built. This is distinct from conventional manufacturing methods, which often rely on removal (by cutting, drilling, chopping, grinding, forging, etc.) instead of addition. Models can be multi-colored to highlight important features, such as tumors, cavities, and vascular tracks. 3DP technology can build a 3D object in almost any shape imaginable as defined in a computer-aided design (CAD) file. It is additive technology as distinct from traditional manufacturing techniques, which are subtractive processes in which material is removed by cutting or drilling [4].

3D printing has started breaking through into the mainstream in recent years, with some models becoming affordable enough for home use. Many industries and professions around the world now use 3D printing. It plays a key role in making companies more competitive. The gap between industry and graduating students can be bridged by including the same cutting-edge tools, such as 3D printing, professionals use every day into the curriculum. There are 3D printed homes, prosthetics, surgical devices, drones, hearing aids, and electric engine components. As shown in Figure 8.1, 3D printing involves three steps [5]. A typical 3D printer is shown in Figure 8.2 [6].

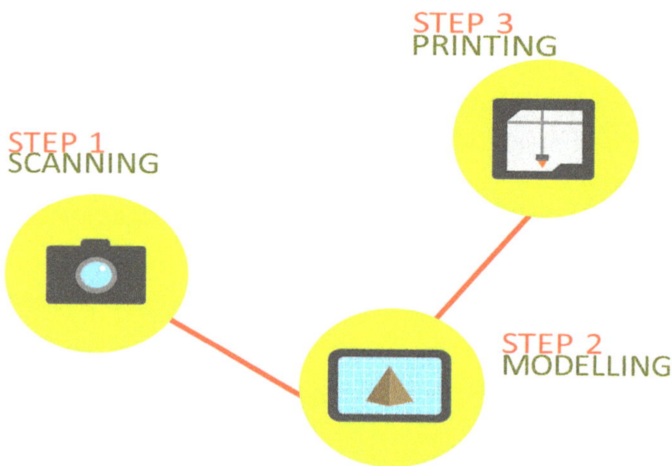

Figure 8.1 3D printing involves three steps [5].

Figure 8.2 A typical 3D printer [6].

8.3 3D PRINTING IN CONSTRUCTION

Although 3D printing has emerged as a promising solution to many of the industry's challenges in recent years, the construction industry has been slow to embrace this new technology. The 3D printing systems created specifically for construction are commonly known as "construction 3D printers" (C3DP). It is the automated process of manufacturing construction elements or entire structures by means of a 3D printer. It usually involves a robotic arm that extrudes concrete while moving back and forth. A typical one is shown in Figure 8.3 [7].

Figure 8.3 A typical construction 3D printers (C3DP) [7].

In construction industry, 3D printing has been used to print houses, offices, and other structures. 3D printers used for construction generally fall into two categories: gantry-style setups and robotic arms, with the first being the dominating solution for on-site printing of entire building structures.

The material used in 3D printed construction is important. The material selected should depend on the type of project in question. The most common materials used are [8]:

Standard concrete,

- Proprietary concrete mixes,
- Mortar,
- Plastic,
- Metal, and
- Local natural materials (stone, sand, mud, rice waste, etc.)

While cement-based 3D printing can create an entire house, it can also be used to create smaller prefabricated construction components. Architects, engineers, and designers can combine 3D printing with other technologies. Figure 8.4 shows the first 3-D-printed, two-story home currently under construction in Houston, Texas [9]. Figure 8.5 shows 3D printing construction workers [10].

Figure 8.4 The first 3-D-printed, two-story home currently under construction in Houston, TX [9].

Figure 8.5 3D printing construction workers [10].

8.4 APPLICATIONS OF 3D PRINTING IN CONSTRUCTION

3D printing in construction is a relatively new technology that replaces some manual work in construction with a machine that creates and assembles structures. 3D printers are used to create building, infrastructure, architectural models, prototypes, formwork solutions, and structural components. 3D printing is used in construction for a variety of applications. These include the following [11-13]:

1. *Buildings:* 3D printing can create entire buildings, including houses and offices, layer by layer. It can help provide shelter to homeless people and shelters for disaster relief. 3D printing can be used to create infrastructure like bridges, railings, and staircases. Because of its near automation, 3D printers can sometimes create houses or other structures within a few days. 3D printing can happen continuously with little supervision. The material used in the 3D printing building is a concrete material reinforced with special glass fiber, and its strength and service life are much higher than that of reinforced concrete. Entire structures can be built more sustainably, with less labor, and in a fraction of the time. A 3D printed building is displayed in Figure 8.6 [13].

Figure 8.6 A 3D printed building [13].

2. *Construction Parts:* 3D printers can create smaller components. They can be used to create construction materials like bricks and modular panels you need from raw materials. 3D printers are useful for creating new structures on-site or manufacturing components off-site for later assembly. The construction teams can produce parts as needed, avoiding waste problems from ordering too many unused pieces during construction. With the improvement of 3D printing technology, more and more construction component can be made by 3D printing.

3. *Creating Prototypes:* Prototypes are preliminary models of structures that give people an example of what something could look like. Prototyping can be an essential part of pre-construction planning. For example, if a company wants to build a new office, a 3D printer could create a prototype for the office much easier than a construction company could level the ground, hire labor, and prepare full construction sites. 3D printing provides clients with the ability to create these prototypes much quicker and at a lower price than constructing full prototypes through traditional building methods. A scale prototype of an entire building helps assess its feasibility and structural integrity. Engineers can test or observe the model to check for potential problem areas.

4. *Architectural Modeling:* The initial manifestation of 3D printing entering the construction industry is architectural models. Modeling is similar to prototyping, except for a few key differences. Modeling is when construction companies or architects scale proposed projects to a size that you can view in a smaller space. This allows clients to inspect the design, layout and other elements of a structure. Architectural modeling during the design stage focuses on the creative and visual aspects of building plans. Modeling can be important for getting investors for a project or showing current stakeholders what the building will look like after completion. With traditional construction, people typically design structures through software and then build models manually. Depending on the size, complexity, and material requirements, 3D printers can quickly produce accurate architectural models. 3D printing can quickly build models once you have created the structure in your computer program and the printer has your materials. It can help make a more accurate representation of the model based on architectural plans.

8.5 BENEFITS

3D printing technology has been a solution that touches almost all areas of the value chain of the construction sector. 3D printers are capable of printing a wide range of architectural models. The construction industry can benefit from 3D printing technology in several ways. Compared to traditional construction methods, 3D printing for construction is generally faster, more efficient, and requires less manual labor. Other benefits of 3D printing in construction include the following [7]:

1. *Flexible Design:* Complex shapes and designs are far more possible through 3D printing than with traditional manufacturing processes. This process has minimal impact on cost and requires less human effort than traditional construction methods. It can help to increase the functionality and aesthetic appeal of structures. Last-minute changes to the design will no longer be a problem nor delay the construction process.

2. *Rapid Production:* 3D printing technology can produce parts

within hours, accelerating the prototyping process. This implies that houses or structures can be built in a few days.

3. *Minimizing Waste:* Construction sites are infamous for creating a lot of waste. However, 3D printing technology uses only the materials needed for each part, resulting in little to no wastage. Because 3D printing creates minimal waste, the work site takes less time to clean up, saving you even more time.

4. *Recycled Materials:* 3D printing can use recycled materials. This benefits both your company and the environment. The construction industry also benefits from recycled concrete and other recycled materials.

5. *Less Human Error:* 3D printing technology can significantly decrease the occurrence of errors during construction. Because of the automated nature of 3D printing technology, the chances of human error, injury, or fatalities are practically eliminated.

6. *Safety:* 3D printing has brought to construction the health and safety of the employees onsite. By knowing how to work effectively with printers, workers can do their jobs more easily and reduce injuries in the field. Another way in which 3D printing enhances safety is by increasing accuracy and precision.

7. *Cost Reduction:* Automation reduces cost of construction. The construction industry is among the least automated/ digitized of all sectors with a major potential for increasing productivity. 3D printing can also reduce the need for heavy machinery, such as cranes and excavators, which can be costly to rent and operate.

8. *Increased Efficiency:* 3D printing can increase efficiency in the construction industry. It does that by significantly reducing construction times. It also increases efficiency is by reducing waste. The reduced wastage and less downtime also result in added cost savings.

9. *Housing Crisis:* 3D printing is considered a very real potential solution to humanitarian issues like the global housing crisis. Since 3D printing now allows structures to be created more quickly, it is ideal to combat the housing crisis. Such an affordable, sustainable housing project in Malawi, Africa is shown in Figure 8.7 [14].

Figure 8.7 An affordable, sustainable housing project in Malawi, Africa [14].

10. *Sustainability:* One of the biggest concerns in all industries today is sustainability and the reduction of CO2 emissions. Sustainability takes center stage with 3D printed houses. As 3D printed constructions involve less waste, these homes are more environmentally friendly. Recycled products can be used to produce the construction materials used in 3D printers.

8.6 CHALLENGES

Although 3D printing technology has shown promise for building large structures, the technology is not yet up to the task. 3DCP faces barriers to unlock its full potential, such as updated building codes/standards and labor upskilling. Other challenges or drawbacks of 3D printing in construction include the following [7]:

1. *Investment Costs:* The cost of 3D printing technology and materials can be high. Buying or even renting a 3D printer

can be very expensive. Many construction professionals find it challenging to justify the cost of 3D printing technology when weighed against its potential benefit.

1. *Labor Shortage:* 3D printing demands specialized set of skills, creating a shortage of skilled workers.

2. *Quality Control:* Ensuring the quality of 3D-printed structures can be a challenge, especially since there are no established industry standards or codes for construction with 3D printing. Since the 3D printing process is automated, construction relies on the printer's accuracy and precision.

3. *Regulation Challenges:* The regulatory framework for 3D printing in the construction realm is currently unclear. This lack of government oversight is a significant obstacle to the rapid adoption of the technology.

4. *Acceptance:* One of the biggest problems in the launch of new technology is the hesitation in acceptance. Would people agree to live in a house built with 3D printing technology? People prefer sticking to traditional methods because of a common misconception that technology makes construction cheaper and weaker.

8.7 CONCLUSION

3D printing is one of the greatest technological advances and innovations of the 21st century. The construction industry is on the brink of a revolution, and at the heart of this transformation lies 3D printing technology. As the technology continues to develop, the possibilities for adoption and growth increase in the construction industry. With the 3DCP market still in its infancy, market size estimates for the industry are yet too premature to trust. Construction giants (such as COBOD, Chinese company WinSun, Italian company WASP, Australian company Luyten 3D) are quickly realizing the potential of 3D technologies and their impact on the future of construction.

3D concrete printing, also known as additive construction, is a promising technology that has the potential to revolutionize the construction of buildings and structures, saving time, materials, labor, and costs, while also enhancing the sustainability and environmental impact of construction. The technology promises to revolutionize how we design and build structures by improving quality, ensuring worker safety, reducing costs, increasing efficiency, and driving sustainable business practices. The use of 3D printing in construction is on the rise and will continue to increase exponentially in the coming years. The "dream" solution is that one day buildings would be printed in concrete, onsite, with robots and gantry systems laying down fast drying cement, working 24/7 to deliver at great speed, affordable structures for housing, offices, barracks, hotels, etc. [15]. More information about 3D printing technology in construction can be found in the books [16-21].

REFERENCE

[1] "Construction 3D printing," *Wikipedia*, the free encyclopedia,

https://en.wikipedia.org/wiki/Construction_3D_printing

[2] M. N. O. Sadiku, P. O. Adebo, and J. O. Sadiku, "3D printing in construction," *International Journal of Trend in Research and Development*, vol. 11, no. 3, May-June 2024, pp. 233-238.

[3] F. R. Ishengoma and T. A. B. Mtaho, "3D printing: Developing countries perspectives computer engineering and applications," *International Journal of Computer Applications*, vol. 104, no. 11, October 2014, pp. 30-34.

[4] M. N. O. Sadiku, S. M. Musa, and O. S. Musa, "3D Printing in the chemical industry," *Invention Journal of Research Technology in Engineering and Management*, vol. 2, no. 2, February 2018, pp. 24-26.

[5] D. Pitukcharoen, "3D printing booklet for beginners,"

https://www.metmuseum.org/-/media/files/blogs/digital-media/3d printingbookletforbeginners.pdf

[6] "The benefits of 3D Printing in children's education," February 2021,

https://www.kidsinthehouse.com/blogs/kidsinthehouse2/the-benefits-of-3d-printing-in-childrens-education

[7] "3D Printing technology for construction,"

https://www.xometry.com/resources/3d-printing/3d-printing-in-construction/

[8] "How 3D printing in the construction industry works," July 2023,

https://www.cemexventures.com/3d-printing-in-construction/

[9] E. Garcia, "3D printing reaches new heights with two-story home," January 2023,

https://www.reuters.com/technology/3d-printing-reaches-new-heights-with-two-story-home-2023-01-12/#:~:text=The%20enormous%20printer%20weighing%20more,square%2Dfoot%20home%20in%20Houston.

[10] C. S. Daugherty, "ERDC researchers are a driving force in construction scale 3D printing," September 2020,

https://www.erdc.usace.army.mil/Media/News-Stories/Article/2356722/erdc-researchers-are-a-driving-force-in-construction-scale-3d-printing/

[11] "3D Printing in construction: types, benefits and uses," August 2024

https://www.indeed.com/career-advice/career-development/3d-printing-construction

[12] "How 3D printing is used in construction," March 2024,

https://www.kreo.net/news-2d-takeoff/how-3d-printing-is-used-in-construction

[13] "Application of 3D Printing House," March 2021,

https://www.eplus3d.com/application-of-3d-printing-house.html

[14] J. Biggs, "6 of the world's most impressive 3d printed buildings," January 2021,

https://www.procore.com/jobsite/6-of-the-worlds-most-impressive-3d-printed-buildings

[15] M. Day, " 3D printing in construction," February 2020,

https://aecmag.com/features/3d-printing-in-construction/

[16] B. Rangel et al. (eds.), *3D Printing for Construction with Alternative Materials (Digital Innovations in Architecture, Engineering and Construction)*. Springer, 2023.

[17] J. G. Sanjayan, A. Nazari, and B. Nematollahi, *3D Concrete Printing Technology: Construction and Building Applications*. Butterworth-Heinemann, 2019.

[18] B. Rangel, A. S. Guimarães, and J. Teixeira (eds.), *3D Printing for Construction in the Transformation of the Building Industry*. Boca Raton, FL: CRC Press, 2024.

[19] A. Perrot, *3D Printing of Concrete: State of the Art and Challenges of the Digital Construction Revolution*. Wiley, 2019.

[20] B. Panda, P. N. Shakor, and V. Laghi(eds.), *Additive Manufacturing for Construction*. Emerald Publishing Limited, 2023.

[21] T. S. Brar, M. A. Kamal, and S. Singh, *3D Concrete Printing Technology Configuration with Green and Self-Healing Concrete*. Materials Research Forum LLC, 2022.

CHAPTER 9
WEARABLES IN CONSTRUCTION

"3 Shallow men believe in luck. Strong men believe in cause and effect."

—Ralph Waldo Emerson

9.1 INTRODUCTION

Technology is advancing at an astounding rate. A wearable is an essential piece of equipment aimed at improving worker safety, health, and productivity. It is a computer-powered device that is never-sleeping ever-present network-connected electronic system that can be used at anytime and anywhere and does not in any way disturb the user's interaction with the real world. It should be worn, much as eyeglasses or clothing are worn, and interact with the user [1]. It includes all manner of technology that is on or in the body such as fitness trackers, smartwatches, smart clothing, smart rings, smart glasses, wearable mobile sensors, smart jewelry, and smart ECG (electrocardiogram) monitors. From Apple Watches to medical alert pendants, wearable technology plays an integral role in our everyday lives. There has been a proliferation of wearables from consumer gadgets to medical devices that are approved by the Food and Drug Administration (FDA). Typical wearable devices are shown in Figure 9.1 [2].

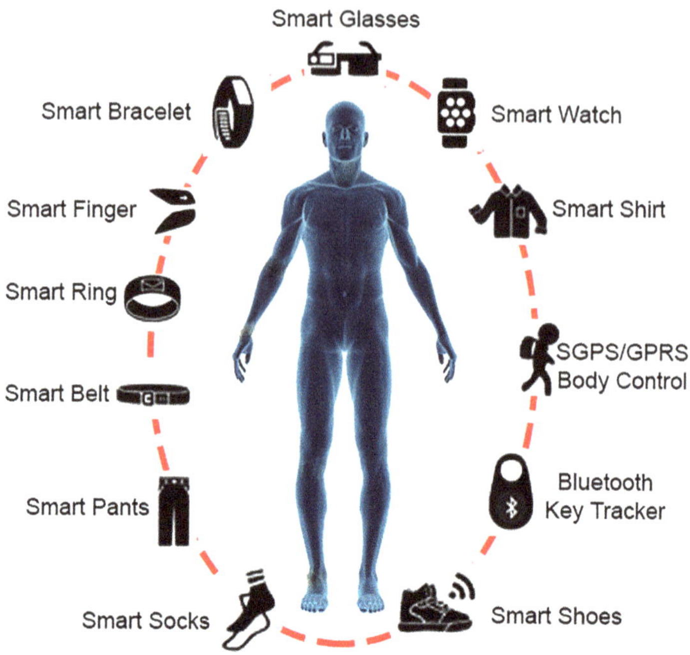

Figure 9.1 Different types of wearable devices on human body [2].

Wearable technology is already revolutionizing many different industries, including construction. Construction has been well known for its labor-intensive nature, hazardous working conditions, and requirement for exact coordination between personnel, equipment, and supplies. If construction companies want to remain relevant, they must invest in new and innovative tools such as wearable technology. Embracing wearable technology and other innovative tools can lead to improved safety, efficiency, and workforce attraction. As the construction industry prioritizes worker safety, wearable technology has emerged as a promising solution. Figure 9.2 shows a typical construction site [3], while Figure 9.3 depicts some construction workers [4].

Figure 9.2 A typical construction site [3].

Figure 9.3 Construction workers [4].

The existing wearable technologies applied in other industrial sectors can be used to monitor and measure a wide variety of safety performance metrics within the construction industry. The use of wearable technology in construction is extensive. Wearable technology is ushering in a new era of construction, from smart helmets with augmented reality (AR). Wearable construction technology refers to any device that construction workers wear on their bodies, to improve safety and reduce injuries [5].

This chapter examines the uses of wearable technologies in the construction industry. It begins with explaining what wearables are all about. It presents different types of wearable devices. It describes wearables in construction and provides some of their applications. It highlights the benefits and challenges of wearables in the construction industry. It concludes with comments.

9.2 BACKGROUND ON WEARABLES

A wearable device is any device that is worn comfortably on the body and enables user interaction. It is typically integrated into the clothing or attached to the body of a person to enhance human performance. It often includes smart devices that can be worn on the body or attached to clothes. Wearable devices have been around for centuries. The first one was introduced in the 1660s by the Qing Dynasty. Since then the popularity of wearables has shifted from royalty to the healthcare industry. Wearable computing is a natural evolution of the smartphone technology that has become so ubiquitous and indispensable in education, business, and medicine. We wear wrist watches to know the time. Perhaps the most crucial bit of wearable tech accessible today is Google Glass. There has been a proliferation of wearables from consumer gadgets to medical devices that are approved by the Food and Drug Administration (FDA).

Wearable devices or systems are usually lightweight, miniature electronic or digital devices that are worn by a user, including clothing, watches, glasses, shoes, and similar items. A wearable computer is computer-powered device that is never-sleeping ever-present network-connected electronic system that can be used at anytime and anywhere and does not in any way disturb the user's interaction with the real world. It should be worn, much as eyeglasses or clothing are worn, and interact with the user [6]. It includes all manner of technology that is on or in the body such as fitness trackers, smartwatches, smart clothing, smart rings, smart glasses, wearable mobile sensors, smart jewelry, and smart ECG (electrocardiogram) monitors.

Wearables are always ready, unrestrictive, not monopolizing of user attention, observable and controllable by the user, attentive to the environment, useful as a communication tool, and personal devices [7]. Wearables are being used across healthcare, insurance, interactive systems, safety critical settings, wearable cameras, baby and pregnancy monitors, entertainment, fitness and sports, emergency responders, and military. Some of these applications are illustrated in Figure 9.4 [8].

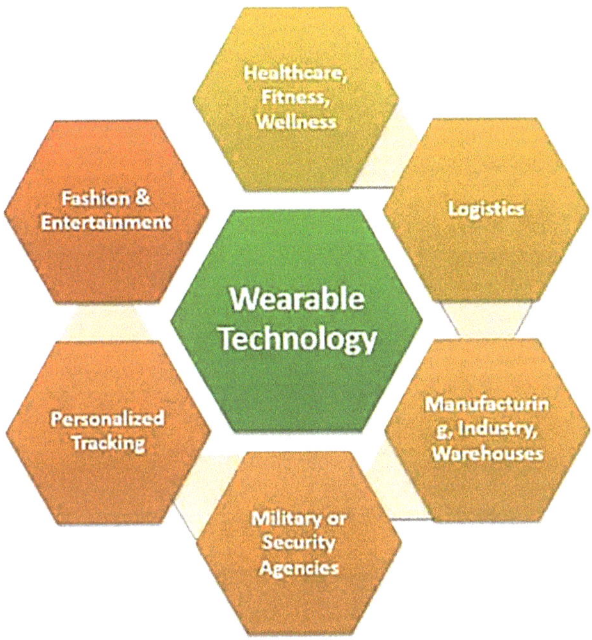

Figure 9.4 Different applications of wearable technology [8].

Due to it wide range of applications, international corporations such Google, Apple, and Intel are investing heavily on wearable technology research and development [9]. As with any new technology, one must exercise caution when using a wearable device.

A wearable device essentially consists of two different components: wearable and body sensors. It incorporates sensors, memory, solar cells, and batteries. It stays in contact with the body for extended periods of time. Traditional materials for wearables are mostly metals and semiconductors with relatively poor mechanical flexibility. Modern wearable technologies are characterized by body-worn devices, as smart clothing, e-textiles, and accessories [10]. Wearable devices can be used to collect various data to support a series of innovative applications. Today, wearable devices have numerous applications due to their integration with artificial intelligence. Wearable devices can be attached to shoes, eyeglasses, earrings, clothing, gloves, and wrist watches [11].

9.3 DIFFERENT TYPES OF WEARABLE DEVICES

The first wearable product to gain massive popularity was fitness bands or activity trackers. Then advanced elements such as a screen, wristwatch, and mobile apps were added. As illustrated in Figure 9.5, AI is making wearables smarter [12].

Figure 9.5 Examples of smart wearables [12].

Wearable devices are used for many purposes such as fitness, monitoring, sleeping disorder, etc. Different types of wearable devices are discussed as follows [13]:

- *Smart Phones:* People usually carry their smart phones around in their pockets. Smart phones have many sensors embedded, which collect data about the movements of users.
- *Smart Watches and Wristbands:* These have various built-in

sensors that monitor users' daily activity, calorie consumption and heart rate, as well as the quality of sleep, etc. Wristbands have recently become a widely used health management tool, with large numbers of people wearing them daily.

- *Smart Glasses:* Smart glasses will not become a threat to privacy, but a practical life assistant and medical tool.
- *Smart Clothes:* Smart clothes collect body data from users, which can be used to monitor users' exercise data and heat consumption. There are also smart baby clothes for infants to monitor their physical condition.
- *Smart Shoes:* Smart sneakers collect users' sports data to help them improve their sports performance.
- *Smart Earphones:* These allow users to operate the equipment more conveniently using voice commands.

There are also smart gloves of manufacturing settings and smart suits for military and space.

9.4 WEARABLES IN CONSTRUCTION

Wearable technology in construction is developing to improve productivity, safety, and data collection, ushering in a new era of effectiveness and creativity. It is transforming the construction sector by improving communication, data collection, safety, and production. Wearables in construction are instruments that boost productivity, empower workers, and raise the bar for safety on construction sites.

Wearable technology in construction is a form of personal protective equipment (PPE) integrating construction safety technology and innovation (e.g., exoskeleton, embedded GPS and telematics, etc.) to make construction work safer and more efficient for the user. It refers to hands-free electronic devices capable of sending and receiving data through the Internet. It plays a crucial role in making sure that construction sites are as safe as possible. Typically worn on the body, clothing, or personal protective equipment (PPE), wearables are used to monitor a worker's environment, activities, and health. From smart watches that monitor vital signs and detect falls, to powered footwear that can track location and call for help, to helmets that detect fatigue and prevent accidents, these wearables are transforming traditional safety equipment into powerful tools that enhance worker safety, health, and productivity. Figure 9.6 illustrates an example of how wearable technology is applied in construction [13]. The hard hat and the vest are equipped with various sensors, ID readers, monitors, GPS tracking, and charging capabilities.

Figure 9.6 How wearable technology is applied in construction [14].

9.5 APPLICATIONS OF WEARABLES IN CONSTRUCTION

There are so many exciting wearables sweeping through the world of construction right now. Common wearables used in construction include the following [15-17]:

1. *Smart Helmets:* Helmets have been standard safety equipment for decades, but now they come with the most advanced features. Smart helmets are used for real-time access to project plans, safety alerts, and communication. In the construction industry, smart helmets are transforming safety and productivity. These helmets have a variety of sensors and functions that give employees and managers access to real-time data. Smart helmets can increase safety. They can keep an eye on environmental factors like noise levels and air quality. They may also track a worker's movements and give information about their level of productivity and weariness. Figure 9.7 displays a typical smart helmet [15].

Figure 9.7 Smart helmet [15].

2. *Smart Hard Hats:* These are a new type of safety wearable designed to provide workers with additional safety and functionality on the job site. They are an excellent solution for workers with long-hour shifts, long-haul truck drivers, and workers who frequently perform monotonous activities. One of the key features of smart hard hats is their ability to detect and alert workers to potential hazards. Coming equipped with advanced technology

including sensors, cameras, and communication systems, this type of equipment can help your team boost its productivity. These can help workers stay safe and connected on the job site.

3. *Smart Gloves:* These aid interaction with digital interfaces, data collection, and documentation. Smart gloves are made with inbuilt touchscreens and sensors to make a variety of jobs on construction sites easier. Without having to handle additional equipment, workers may use these gloves to communicate with digital interfaces, take measurements, and record their progress.

4. *Smart Clothing:* This is used for continuous monitoring of vital signs and posture. Traditional work gear can be equipped with sensors and electronics to improve safety and enable real-time health monitoring. These devices can keep an eye on critical indications including posture, body temperature, and heart rate. Smart clothing in the construction industry can warn employees and managers of potential health concerns or weariness, improving wellbeing at work.

5. *Smart Boots:* There may be no piece of equipment that a construction worker relies on as much as their work boots. These are used to tracks gait, posture, and movement patterns. The gait, posture, and movement patterns of workers are tracked by sensors built into smart boots. The analysis of ergonomics and the detection of potential musculoskeletal concerns can be done using this information. Smart boots give managers information on how physically fit their employees are, allowing them to improve workflow and avoid accidents caused by poor posture or movement. Along with their advanced technology, smart boots also improve on comfort and efficiency features. By harnessing the power generated from walking, these boots can be charged throughout a worker's shift without needing external charging stations. Typical smart boots are shown in Figure 9.8 [15].

Figure 9.8 Typical smart boots [15].

6. *Smart Glasses:* Designed to enhance performance, smart glasses are changing the way construction workers perform their tasks. They provide real-time remote support and documentation of construction tasks. Smart glasses with cameras and augmented reality (AR) technologies are being employed in the building industry. These glasses can be worn by employees to record real-time photos or videos of construction activities. Figure 9.9 shows some typical smart glasses [4].

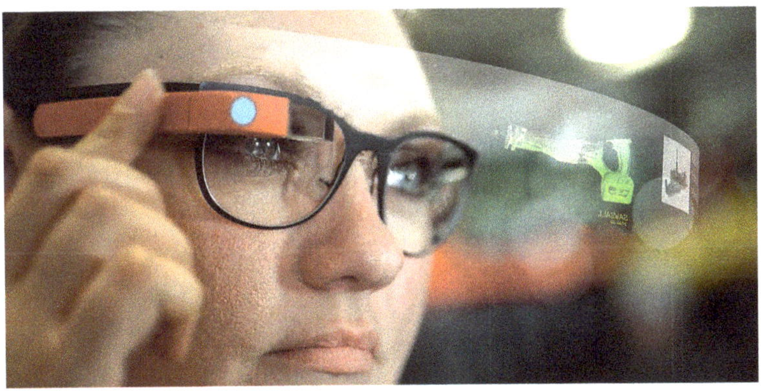

Figure 9.9 Typical smart glasses [4].

7. *Smart Watches:* Smart watches, which many people commonly wear daily, have become increasingly relevant in the construction industry due to their powerful sensors. They are known for their ability to track physical activity and heart rate monitoring.

However, they can also be a valuable tool for construction workers, offering a variety of features that benefits employees on the job site like being hands-free. These can help you measure pertinent health data and prevent overexertion. The most commonly used feature of smart watches is their health monitoring system, which tracks heart rate, steps, calories burned, and other metrics. This can let workers/users know when their heart rate is too high and they need to take a break. Smart watches can also be beneficial for safety and communication. They allow workers to stay connected to their colleagues and receive updates and notifications in real-time. Figure 9.10 displays a typical smart watch [17].

Figure 9.10 A typical smart watch [17].

8. *Smart Safety Vests:* Safety vests are another staple on the construction site, but now they come with high-tech features. They play a key role in keeping workers safe when they enter off-limits hazard zones, like the space behind a piece of moving heavy machinery. If employees come into these areas, the vest will light up and send an alarm. Many safety vests are now equipped with sensors that can be clipped on or embedded into the clothing. Similar to other wearable tech, safety vests can be used to measure workers' physiological conditions and alert them of job site dangers. Because safety vests are close to the wearer's body, you can accurately track different biometrics such as body temperature, pulse rate, and blood pressure.

9. *Exoskeletons:* An exoskeleton is a wearable, biomechanical suit or apparatus that enhances the physical attributes and abilities of the person wearing it. It assists workers with physically demanding tasks like lifting and reduces strain. The use of exoskeletons, gadgets such as wearable technology in construction that give employees physical support, is growing in the construction sector. By increasing a worker's strength, these tools help lessen the physical stress that comes with jobs like lifting large objects or running powerful machinery. Exoskeletons can also have a big impact on construction projects by making labor-intensive tasks more effective. Less effort is required for workers to complete their tasks, which could shorten project schedules and save labor costs. Modern exoskeletons fall into two overarching categories: partial and full-body. A construction exoskeleton can take many different forms depending on the needs of the job at hand. Exoskeletons have the potential to dramatically increase the health and safety of workers. Figure 9.11 displays a modern exoskeleton [4].

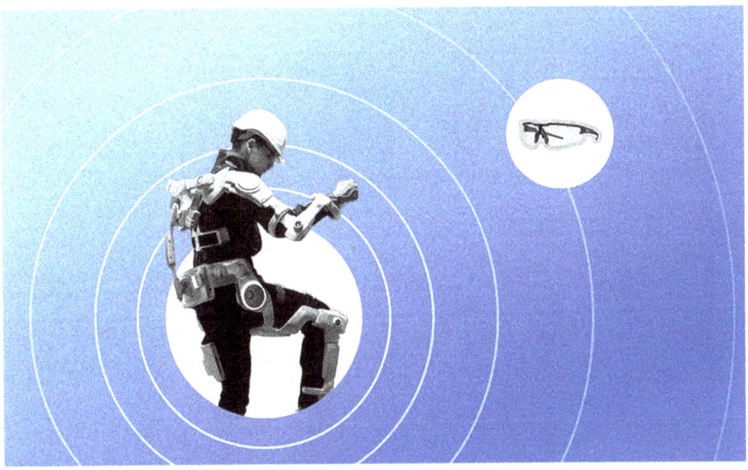

Figure 9.11 A modern exoskeleton [4].

10. *Wearable Robotics (Exosuits):* These assist with lifting heavy loads and physically demanding tasks. Wearable exosuits, commonly referred to as soft exoskeletons, are intended to let construction workers handle large objects and carry out strenuous jobs more easily. The power and endurance of a worker can be increased by these wearable robotics, making it simpler to handle

heavy equipment and materials.

11. *Augmented Reality (AR) Glasses:* Augmented reality (AR) glasses offer an innovative approach to construction safety by overlaying computer-generated imagery onto the real-world environment. These glasses provide workers with additional layers of information, enhancing their situational awareness and improving safety protocols. They help in visualizing building plans, providing on-site guidance, and improving task accuracy. AR glasses are becoming more and more common in construction. These wearable gadgets give workers immediate access to digital data and visualizations that are superimposed on their physical surroundings. With AR glasses, construction workers may visualize building designs, receive on-site coaching, and see structures that are concealed within walls or floors. Figure 9.12 displays a construction worker with AR glasses [17].

Figure 9.12 A construction worker with AR glasses [17].

12. *GPS-Enabled Wearables:* These wearables are used for real-time worker location tracking and site logistics optimization. On construction sites, GPS-enabled wearable gadgets are utilized to monitor the whereabouts of workers. By giving managers the ability to track employees' whereabouts in real-time, these gadgets can increase safety and help to optimize work processes.

13. *RFID Wearables:* These are used for tracking asset locations,

usage, and maintenance. Wearables with RFID (Radio-Frequency Identification) technology are better able to track and manage assets on construction sites. Equipment and supplies can be marked using RFID chips, and employees can wear RFID wristbands or tags. This makes it possible to track assets' whereabouts, uses, and maintenance statuses effectively in real-time.

9.6 BENEFITS

By solving safety concerns, boosting productivity, enabling data-driven decision-making, and increasing communication and collaboration, wearable technology is gradually changing the construction sector. The construction industry can gain a lot from the successful integration of wearable technology in construction, increasing its competitiveness, safety, and efficiency. For all the exciting benefits these new tools bring, it is important to remember that their truest value derives from the extent to which they improve the world and the lives of the people who use them. Other benefits include the following [15]:

- *Ease of Use:* Wearable technology has evolved and become more sophisticated since its introduction. These portable and hands-free gadgets/devices are easy to set up and connect with smartphones. They can be worn anywhere and are available in multiple forms such as wristwatches, accessories, glasses, headsets, etc.

- *Enhancing Safety:* One of the biggest benefits of wearable technology in construction is the safety improvements. The high rate of fatalities in the construction industry remains a major concern of both practitioners and researchers. Therefore, in the construction industry, safety is a top priority, with a constant focus on reducing injuries and fatalities. Wearable devices are typically worn on a worker's waist belt, allowing for constant connection with site supervisors. Smart helmets, for example, can detect threats like falling items or high temperatures and warn users of them. The emphasis on safety in the construction sector is one of the main factors driving the adoption of wearable technology. Wearables like smart helmets that have sensors and cameras incorporated in can track a worker's movements and the surrounding surroundings.

These tools assist in the reduction of accidents and injuries. If an accident occurs, wearable technologies can help to save lives through slip and fall detection. Through GPS technology, it can alert supervisors and give them a worker's exact location.

- *Boosting Productivity:* Wearable technology in construction can greatly increase worker productivity by giving them immediate access to vital information. For example, augmented reality (AR) glasses project digital information into the actual world, helping workers with tasks like map reading, equipment operation, and navigation. This technology reduces mistakes and expedites work.

- *Reduced Costs:* Wearables enable remote collaboration by making data easily available, ensuring that you save time and resources. In most cases, wearables feed data to customized dashboards, which enables stakeholders and supervisors to view data and draw insights with specific job sites, job categories, job functions, and employees, thereby reducing rework and the costs.

- *Data Collection:* The wearables collect data about the workers and the worksites. Massive volumes of data are collected by wearable technology in construction, and this information is useful for decision-making and project management. Wearable technology streamlines collection process, giving you a more accurate and thorough set of data. Project managers can streamline workflows and guarantee employee well-being by using wearable sensors to monitor workers' movements and vital signs. Data on worker behaviors and worksite circumstances can be collected via wearable technology in construction. For example, wearable GPS trackers can deliver accurate location information for employees and equipment, facilitating better resource management.

- *Collaboration:* Wearables enable construction teams to communicate and collaborate more effectively. Without the use of paper documents or mobile devices, this technology enables workers to share information, view plans, and get instructions, enhancing teamwork and lowering error rates. This improves problem-solving and teamwork on building sites.

- *Communication:* Real-time communication between

construction teams is made possible by wearable communication equipment like smart glasses or headsets. Employees who have smartwatches or headsets at their disposal can communicate in real-time, share information, and get project updates without stopping what they are doing. Gone are the days of chasing down other workers or playing phone tag. Wearable technology keeps you connected to your team members at all times and ensures that everyone is kept in the loop.

- *Productivity:* Wearable technology in construction is being used to increase productivity. By supporting heavy lifting and repeated jobs, exoskeletons, for example, can lessen the physical stress on construction workers. This increases worker comfort and productivity, while lowering the risk of musculoskeletal injuries.

- *Health:* Wearable technology is becoming more and more popular with people of all ages. It allows individuals, with age no bar, to track their vitals as well as creates a more connected health experience for both healthcare seekers and givers. It allows individuals to track their vitals as well as creates a more connected health experience for both healthcare seekers and givers. From fitness trackers to smartwatches, wearable technology helps people stay healthy, organized, and in control of their lives.

- *Biometric Monitoring:* Monitoring can help tracks vital signs and alerts workers to health risks. In order to monitor employee wellbeing in the construction business, biometric monitoring wearable technology in construction is becoming more and more crucial. Real-time monitoring of vital signs including heart rate, body temperature, and hydration levels is possible with these gadgets. Accumulating biometric data over time can offer important insights into the long-term health of construction workers.

- *Standards:* Standards may have improved vastly, yet they vary considerably across the globe. Wearable technology has the potential to transform safety standards in the construction sector on top of improving efficiency and productivity.

9.7 CHALLENGES

Although wearable technology in construction has a lot of potential, adoption in the construction industry faces some challenges, including worker skepticism, employer reservations, and technical and logistical complexities. Aspects including costs, data security worries, and the requirement for employee training must be taken into account. Cost, data security, and user acceptance issues must be resolved when the construction sector adopts wearable technology. One common challenge with wearable technology is the need for regular charging. Other challenges include the following [15]:

- *Cost:* A crucial aspect of adopting wearable safety technology is conducting a thorough cost-benefit analysis. The expense of purchasing and maintaining wearable technology in construction is one of the key obstacles. Although these technologies have many advantages, the initial cost can be high. To justify the cost, construction businesses must carefully assess the return on investment (ROI).
- *Data Privacy:* Wearable technology in construction gathers private information, such as worker biometrics and location data. It is essential to safeguard this data against theft or misuse.
- *Data Security:* Security features and provisions for various wearable devices might differ. Therefore, a user should go through the security features of each device carefully in order to protect the personal data collected and shared. The more the technology is widely used, the more attraction it gets from hackers and fraudsters.
- *Data Overload:* This may result from the massive volumes of data that wearables collect. For businesses to get useful insights, appropriate data management and analysis processes must be in place.
- *Integration with Existing Systems:* It might be challenging to integrate wearable technology in construction with current construction management systems and procedures.

Compatibility problems could occur.

- *Safety:* Worker safety cannot be compromised by wearable technology in construction. They should be built to withstand the physical impact, dust, and debris present on construction sites. To prevent equipment failures that could be dangerous; reliability is essential.
- *Regulatory Compliance:* While regulations have helped decrease accidents over the years, there is still room for improvement. It is crucial to adhere to industry-specific rules and safety requirements. Employers in the construction industry must make sure that their wearable equipment complies with these standards.
- *Cultural Change:* The organization may need to undergo a culture transition in order to implement wearable technology in construction. Instead of using technology as a means of monitoring employees, workers should embrace it as a tool that enriches their work.
- *Scalability:* As projects grow or new technologies appear, businesses must take into account how easily wearable technology in construction can scale.
- *Limited Power:* It should be noted that wearable devices require power to work and the power is limited. Wearable devices have a limited battery capacity which may not be a favorable feature for a user or a business that highly relies on devices. Good quality battery devices come at a higher cost.

9.8 CONCLUSION

Wearable technology is rapidly advancing in the construction industry, where efficiency, safety, and data-driven decision-making are critical. It will become more common on construction sites due to its great advantages in improving worker safety and efficiency. Smart wearables are already becoming an essential piece of equipment, aimed at improving worker safety, health and productivity. As the construction industry continues its digitization journey, so does its rapid adoption of wearables.

Wearable technologies have created a paradigm shift in the way we use technology. The future of safety in construction involves powerful wearable technologies. The future of safe construction involves user-friendly and powerful wearable technology, including smart watches, boots, helmets, glasses and body wear. Each tool is designed to keep workers safe, healthy, and productive.

REFERENCE

[1] M. N. O. Sadiku, S. Alam, and S. M. Musa, "Wearable computing," *International Journal of Engineering Research*, vol. 6, no. 10, Oct. 2017, pp. 445-447.

https://www.researchgate.net/figure/Different-types-of-wearable-technology_fig5_322261039

[2] "How the modular industry uses wearable technology in construction," Unknown Source

[3] M. Anderson, "Wearable tech & safety: How exoskeletons and smart glasses make construction safer," June 2024,

https://onekeyresources.milwaukeetool.com/en/2-construction-wearables-explored

[4] M. N. O. Sadiku, C. M. M. Kotteti, and J. O. Sadiku, "Wearables in the construction industry," *International Journal of Trend in Research and Development*, vol. 11, no. 6, November-December 2024, pp. 244-250.

[5] M. N. O. Sadiku, S. Alam, and S. M. Musa, "Wearable computing," *International Journal of Engineering Research*, vol. 6, no. 10, Oct. 2017, pp. 445-447.

[6] M. Kaiiali, "Designing a VM-level vertical scalability service in current cloud platforms: A new hope for wearable computers," *Turkish Journal of Electrical Engineering & Computer Sciences*, vol. 25, 2017, pp. 2555 – 2566.

[7] Canopus Infosystems, "What are the latest trends in wearable technology?"

https://yourstory.com/mystory/3073273d91-what-are-the-latest-trends-in-wearable-technology-

[8] M. Salahuddin and L. Romeo, "Wearable technology: Are product developers meeting consumer's needs?" *International Journal of Fashion Design, Technology and Education*, vol. 13,

no. 1,2020, pp. 58-67.

[9] M. N. O. Sadiku, P. O. Adebo, A. Ajayi-Majebi, and S.M. Musa, "Wearable healthcare technologies," *International Journal of Trend in Research and Development*, vol. 7, no. 3, May-June 2020, pp. 94-97.

[10] M. N. O. Sadiku, O. D. Olaleye, A. Ajayi-Majebi, and S. M. Musa, "Wearable AI: A primer," *International Journal of Trend in Research and Development*, vol. 8, no. 1, Jan.-Feb. 2021, pp. 35-38.

[11] P. Kedia, "How to use AI to enhance today's wearables," September 2019,

https://developer.qualcomm.com/blog/how-use-ai-enhance-today-s-wearables

[12] C. Y. Jin, "A review of AI technologies for wearable devices," *IOP Conference Series: Materials Science and Engineering*, 2019.

[14] "The case for wearables in construction,"

https://www.oracle.com/webfolder/technetwork/tutorials/jdedwards/White%20Papers/Wearables-in-Construction.pdf

[13] "Top 12 wearable technology in construction industry," November 2023,

https://neuroject.com/wearable-technology-in-construction/

[14] T. Forestell, "The best wearable technologies for construction workers," December 2022,

https://dozr.com/blog/wearable-technologies-in-construction

[15] "The emergence of wearable construction technology," October 2023,

https://abcsocal.org/wearable-construction-technology/

CHAPTER 10
AUGMENTED REALITY IN CONSTRUCTION

"Augmented reality is the future of smart construction, and we are just starting to see the possibilities"

—Anonymous

10.1 INTRODUCTION

Augmented reality (AR) is an immersive technology that overlays digital information, such as images, videos, or 3D models, onto the real-world environment in real-time. It combines gaming technology with wearable technologies like smart glasses and headsets. It allows for the creation of 3D models and visualizations that can be overlaid with a physical environment. It has been gaining momentum in various industries over the past few years, and the construction industry is no exception.

Since the 1940s, there have been very few technological changes in the construction industry. The industry is slowly but steadily entering the market of mobile applications. Today, the construction industry is undergoing a digital transformation, and augmented reality (AR) is at the forefront. The last few years have seen a tremendous rise in the usage of augmented reality in different fields, the reason being its ability to allow a user to view fabricated real-world scenarios by infusing their minds with computer-generated visuals such as audio, video, images, etc. As in many other industries, there is a lot of machinery involved in building. Most of it requires hours of training. With the help of an AR headset, an employee can receive direct instruction and act accordingly. As a result, the AR training costs lower, the instructions are more intuitive, and the downtime is minimized. When AR is used in construction, it can provide workers with a safer training environment. Figure 10.1 shows a typical construction site [1], while Figure 10.2 shows some construction workers [2].

Figure 10.1 A typical construction site [1].

Figure 10.2 Some construction workers [2].

Augmented reality (AR) refers to technologies that bring computer-generated objects such as images, sounds, texts, or other digital information into the user's real-life surroundings. It is a technology that superimposes digital information onto the real world. The construction industry has adopted AR technology to improve the design, planning, and construction processes. AR technology allows architects and designers to visualize 3D models of their designs in a real-world environment. It has the potential to revolutionize the construction industry [3].

In this chapter, we will discuss how augmented reality is used in construction. It begins with presenting different types of digital realities. It describes what augmented reality is all about. It discusses augmented in construction and provides some of its applications. It highlights the benefits and challenges of augmented reality in the construction industry. It concludes with comments.

10.2 TYPES OF DIGITAL REALITIES

Architects and builders have always relied on visual representations to convey their designs before beginning construction. The potential uses of AR for client engagement in architecture and construction are extensive. Augmented reality (AR) and its cousin, virtual reality (VR), have revolutionized practices in important fields such as medicine, education, and military training, and they are also changing what is possible in the fields of architecture and construction. Whereas virtual reality is strictly a digital experience, augmented reality combines the physical and digital world into one immersive environment. Unlike virtual reality, which creates a totally new and independent environment of the real world, augmented reality includes virtual elements that interact with what already exists.

In general, there are four types of digital realities [4,5]:

- *Augmented reality (AR)*— designed to add digital elements over real-world views with limited interaction.
- *Virtual reality (VR)*— immersive experiences helping to isolate users from the real world, usually via a headset device and headphones designed for such activities.
- *Mixed reality (MR)*— combining AR and VR elements so that digital objects can interact with the real world means businesses can design elements anchored within a real

environment.

- *Extended reality (XR)*— covering all types of technologies that enhance our senses, including the three types previously mentioned.

These four types are illustrated in Figure 10.3 [6]. As shown in Figure 10.4, the four types form the mixed reality continuum that captures all possible combinations of the real and virtual worlds [7]. AR effectively "augments" or "enhances" our experience of the world around us.

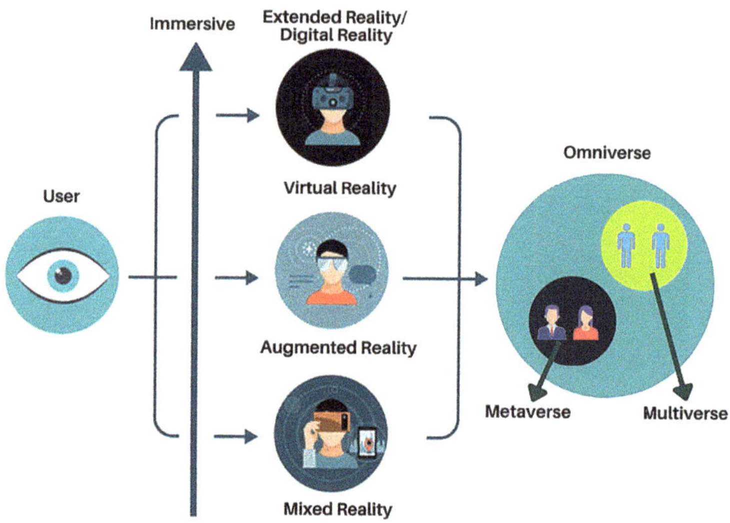

Figure 10.3 Four types of digital realities [6].

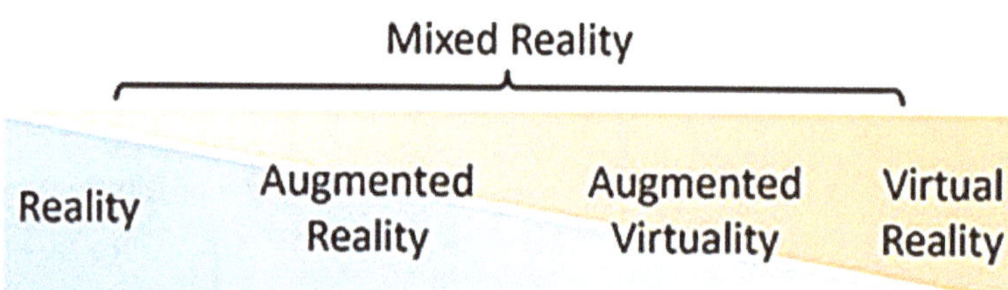

Figure 10.4 The mixed reality continuum captures all possible combinations of the real and virtual worlds [7].

10.3 WHAT IS AUGMENTED REALITY?

The term "augmented reality" was coined in 1990 by Tom Caudell, a former Boeing researcher. Since then, the technology has vastly improved. The rise in AR use stems from four main developments [8]: (1) the pervasiveness of low-cost visual sensors, such as phone cameras, (2) progress in environmental perception algorithms, such as visual simultaneous localization and mapping, (3) advances in optics, and (4) the maturity of multimedia techniques.

Augmented reality (AR) is a technology that combines real-world environments with computer-generated generated information such as images, text, videos, animations, and sound. It can record and analyze the environment in real-time. The technology is accessible to the ordinary user. It is becoming more attractive as a mainstream technology due to the proliferation of modern mobile computing devices like smartphones and tablet computers with location-based services. For example, AR allows consumers to visualize a product in more detail before they purchase it. This feature enhances consumer interaction and helps them never to repurchase the wrong item.

The key objective of AR is to bring computer-generated objects into the real world and allows the user only to see them. In other words, we use AR to track the position and orientation of the user's head to enhance/augment their perception of the world. Projected images are overlaid on top of a pair of goggles or glasses, which allow the images and interactive virtual objects to lay on top of the user's view of the real world. Thus, augmented/enhanced reality involves extending the real-time environment with a digital overlay, in which computer-generated images and sound enhance real life. The digital overlay scene appears on the actual scene the user is experiencing.

Augmented reality falls into two categories: 2D information overlays and 3D presentations, like those used with games. It combines multiple technologies allowing users to interact with virtual entities in real-time. Various technologies used for augmented reality include a processor, monitors, handheld devices, display systems, sensors, and input devices. Modern mobile computing devices like smartphones and tablet computers contain these elements also, making them suitable AR platforms.

To obtain a sufficiently accurate representation of reality, AR needs the following five components [9]:

- *Sensors:* AR needs suitable sensors in the environment and possibly on a user, including fine-grained geolocation and image recognition. These are activating elements that trigger the display of virtual information.
- *Image augmentation:* This requires techniques such as image processing and face recognition.
- *Head-mounted Display:* HMDs are used to view the augmented world where the virtual computer-generated information is properly aligned with the real world. Display technologies are of two types: video display and optical see-through display. AR handset is displayed in Figure 10.5 [10].

Figure 10.5 AR handset [10].

- *User Interface:* This includes technologies for input modalities that include gaze tracking, touch, and gesture. AR is a user interface technology in which a camera-recorded view of the real world is augmented with computer-generated content such as graphics, animations, and 2D or 3D models.

- *Information infrastructure:* AR requires significant computing and communications infrastructure undergirding all these technologies. The infrastructure determines what real-world components to augment, with what, and when.

The AR systems, based on these technologies, should be more accurate, smaller, lighter, faster, simpler, cheaper, and convenient for the users. Google Glass was the first AR platform to get broad public exposure. Figure 10.6 shows the evolution of AR [11].

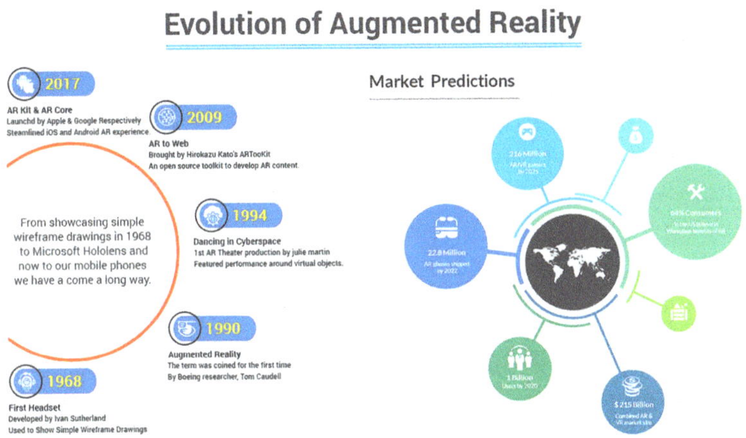

Figure 10.6 The evolution of AR [12].

AR technology has been used in many fields, such as education, healthcare, military, business, engineering, architecture, robotics, manufacturing, entertainment, space industry, maintenance, coding, consumer design, psychology, etc. When incorporating AR in the classroom, it is important to keep the following best practices in mind to ensure a smooth and effective integration [12].

Setting clear goals and objectives for using AR;

- Creating a plan for how AR will be used in the classroom;
- Providing training for educators and students on how to use AR tools;
- Incorporating AR into existing curriculum and lesson plans;
- Assessing and evaluating the effectiveness of AR in achieving

educational goals.

10.4 AUGMENTED REALITY IN CONSTRUCTION

The construction industry is one of the oldest and most important industries in the world. But it is slow in embracing technology and has remained relatively unchanged. Over the last decade, numerous technologies have been introduced into the industry. One of the most exciting technologies in modern construction is the use of augmented reality (AR). Augmented reality essentially uses an advanced camera and sensor technology to create an enhanced version of the real world by adding digital visual elements, sound, or other sensory stimuli and presenting it in real-time. AR technology can revolutionize how construction teams work, from design visualization and collaboration to on-site construction assistance and improved long-term maintenance. It has transformed the way we build and design structures and has become an essential tool for architects, engineers, and construction professionals.

In the construction industry, augmented reality technology has been observed to be utilized for five main purposes. These purposes are: improving project modelling, enhancing safety measures, providing better education opportunities, enabling more effective inspections, and facilitating construction assemblies [13]. Figure 10.7 shows the use of AR in construction [14].

Figure 10.7 Use of AR in construction [14].

10.5 APPLICATIONS OF AR IN CONSTRUCTION

AR overlays digital information such as images on the real world and in real time. It has been creating waves in the construction industry due to the wide range of applications it provides throughout a project's lifecycle. It is being utilized in the construction industry to enhance efficiency, improve safety in construction sites, optimize teamwork and collaboration and manage time, money and resources thanks to its ability to deliver real-time information. Common areas of applications of AR in construction include the following [15]:

- *On-site Construction:* Every contractor knows the frustration of arriving on-site, only to discover something is not where it should be. AR will accelerate construction processes by allowing team members to have quick and instant access to virtual models of the project. AR technology can significantly impact site analysis, changing how professionals assess and plan construction projects. AR technology can be used on-site to improve construction accuracy and efficiency. By overlaying digital models in the real world, construction workers can ensure that the building components are installed in the correct locations and at the right angles. This can help to reduce errors and rework, saving time and money. Companies will also be able to benefit from prefabricated components or parts that can be built based off of the virtual models.

- *Safety Training:* Safety is a common challenge in the construction industry. Safety programs cost a lot of time and money and are not always properly communicated to workers. For the actual safety training, it would put the equipment in the hands of trainees as helpful training media is superimposed over the physical world. There is no substitute for learning on the spot, using quality tools. AR can also be used in the construction sector to provide immersive and interactive training experiences for safety procedures, protocols, and awareness. With AR headsets, workers can practice safety procedures in a simulated environment, making it easier to identify potential hazards and learn how to respond to them.

• *Maintenance:* AR technology can be used for building maintenance and repair. By overlaying digital information onto physical objects, maintenance workers can quickly identify equipment and access information about maintenance procedures. This can help reduce downtime and improve maintenance efficiency. Augmented reality adds long-term benefits to building owners and operators, which can help construction companies close more deals. Maintenance workers can also use AR to access digital manuals and repair guides.

• *Underground Construction:* When new cables or pipes are laid, there is always a risk of hitting a gas line or damaging underground utilities. Using augmented reality, you can have all the critical information at your fingertips at any time (depth, ground and cable types, etc.). In the excavation process, there is always a risk of potentially hitting a gas line or wrecking underground utilities. One wrong knick of an excavator's bucket can cause catastrophic explosions or harm. Figure 10.8 depicts an underground construction [16].

Figure 10.8 Underground construction [16].

• *Project Planning:* The value of augmented reality in construction begins in the planning and permitting stages. In the planning phase of construction, it is crucial to make sure that your project plan aligns with the requirements of the client. The use of augmented reality gives construction companies the ability to understand the building more in-depth. AR can display the environmental and social impacts of the project visually. Morpholio is an example of an app that can be used by architects to design the project by incorporating CAD in the planning phase.

- *Project Management:* The real-time visualizations with AR take all the guesswork out of the picture because there is no longer a need to rely on manually recorded measurements or drawings. AR will display accurate images, re-creating entire spaces. As a result, planning and managing projects will become much more efficient and save time for the whole team.

- *Design Analysis:* With AR, project managers will be able to accurately analyze the design because they can walk through the entire project and correct any errors with just a few taps of the screen. Typically, inspections of this kind only took place after the building's construction, when the tiniest mistake could cost the company thousands of dollars.

10.6 BENEFITS

Augmented reality provides a range of benefits that can help companies create the world of tomorrow more effectively and efficiently. This powerful tool enhances accuracy, helping teams catch errors early, prevent rework, and keep projects on track. AR's ability to overlay digital information onto the real world promises to enhance productivity, reduce costs, and improve collaboration in a wide range of construction tasks. Other benefits of AR technology in the construction industry include [17]:

- *Design Visualization:* AR in construction provides a realistic and immersive experience, allowing architects, engineers, and builders to see how the building will look in its intended environment. AR technology can show real-time measurements and other information at the future project site. This data overlaid in the real world can be used to identify potential design flaws before construction begins. AR allows workers to visualize and discuss design plans, share measurements, and track progress. It also allows architects and designers to visualize 3D models of their designs in a real-world environment, making it easier to identify potential design flaws.

- *Safety:* Perhaps the most significant benefit of bringing augmented reality technology into the construction industry is its potential to improve workplace safety. The construction industry is subject to strict regulations and compliance requirements. The Occupational Health and Safety Administration of the United States (OSHA), like many of the governing bodies of other countries, enforces safety compliance. Augmented reality can play a significant role in enhancing safety in the construction industry. Construction is one of the most dangerous industries in fatality rate and days lost to injuries. Safety starts with training, and augmented reality provides realistic training for safety scenarios. AR technology can be used to train workers on safety procedures, reducing the risk of accidents and injuries.

- *Increased Efficiency:* Human errors are a fairly common cause of production disruptions and as a result of money losses. One of the reasons for the mistakes is the low qualification of construction staff. AR technology can help to reduce errors and rework, saving time and money. AR will allow workers to perform any task properly in a safe environment and minimize staff errors.

- *Collaboration:* Construction is a highly collaborative sector, and effective teamwork between all the different disciplines is imperative to ensure that a project is finished on schedule and under budget. Any team's success depends on how well its members work together. AR solutions will facilitate collaboration between different teams on the jobsite, improving communication, coordination, and decision-making. AR can revolutionize the construction industry by providing real-time data and visualization of construction projects in a collaborative and controlled way.

- *Communication:* Augmented reality can share information and data in real-time, allowing professionals from different disciplines to work together more effectively. Miscommunication and project management errors can result in higher costs and missed deadlines, which may lead stakeholders to lose confidence. AR can enhance client-contractor communication. Rather than interpreting complex architectural drawings, clients could walk through a virtual model of their project, ensuring they have a clear understanding and are satisfied with the proposed design. A walk-through of the AR model can give clients a realistic view of the final construction, ensuring expectations align with the planned deliverables. By enabling a shared visual experience that improves efficiency, communication, and understanding, augmented reality has the potential to drastically improve communication and collaboration.

- *Improved Accuracy:* The idea is that AR applications can provide a more accurate view of what will be built, including all layers of materials and installations that are often complex to understand through drawings. AR technology opens the door to the kind of precision, efficiency, and real-time problem-solving that the old methods just cannot keep up with. It can help to improve the accuracy of construction projects by ensuring that building components are installed in the right locations.

- *Cost Savings:* AR can increase labor productivity, reduce costs, and improve safety by using it throughout the project lifecycle.

10.7 CHALLENGES

While AR technology has made significant strides in recent years, its effect may have specific limitations worthy of consideration. Like many emerging technologies, augmented reality is not perfect and has lots of room for improvement and optimization. Comfort and safety are major concerns for users wearing AR headsets for long working hours. AR headsets can be uncomfortable for extended periods, especially in extreme temperatures. They have the potential to be distracting for workers. Not all augmented reality glasses allow their wearer to have a hard hat on while using the device. Other challenges include [10,14]:

- *Data Security and Privacy:* As AR technology becomes increasingly integrated into the construction industry, data security and privacy will become a critical concern for both construction companies and AR developers. Sensitive project information and proprietary company data will all be handled through AR applications, making it essential to protect this data from unauthorized access, misuse, and breaches. To address these concerns, AR product developers will start incorporating robust data security and privacy measures into their applications at an earlier stage. AR systems collect and process sensitive data about the construction site, including worker locations and project details. Implement robust data security measures, including encryption and access control, to protect sensitive information.

- *Cost:* The application of augmented reality (AR) in the construction sector is an innovative move that has significant economic implications. A cost-benefit analysis can help provide a comprehensive understanding of its financial impact. One of the main factors that hinder the mass implementation of AR in construction is the high price of augmented reality. Hence, only large AEC enterprises can afford large up-front investments in emerging technologies like AR. Safety programs take a lot of time and money to implement, and they aren't usually well conveyed to employees. Implementation of augmented reality in construction entails investment in AR hardware and software, staff training, and regular maintenance.

- *Integration:* AR systems need to seamlessly integrate with existing construction software and data systems. Choose AR platforms that offer robust data integration capabilities and compatibility with common construction software.

- *Limited Accessibility.* AR devices, especially those with high-resolution displays and processing power, can be expensive and bulky. Start with AR-enabled smartphones or tablets, which are more readily available. As your requirements grow, consider investing in dedicated AR headsets for more immersive experiences.

- *Regulatory Compliance*: As construction companies adopt AR technologies, it is crucial to navigate and comply with regulatory frameworks and address ethical considerations. Ensure that the implementation of AR complies with relevant industry regulations and standards. Stay informed about any legal or regulatory changes that may impact the use of AR in construction. For international projects, comply with export restrictions and ensure that the use of AR tools aligns with the laws of the countries involved in the construction project.

- *Accessibility Compliance:* Consider accessibility requirements to ensure that AR tools are usable by individuals with disabilities. Compliance with accessibility standards will promote inclusivity and prevent potential legal issues.

- *Intellectual Property Rights:* Construction projects often involve various stakeholders, each contributing to the design and implementation. Respect intellectual property rights and ensure that the use of AR technology does not infringe on patents, copyrights, or other proprietary rights.

- *Ethical Considerations:* Handle data ethically, avoiding the misuse or unauthorized sharing of information collected through AR. Obtain informed consent from individuals whose data may be captured or processed by AR systems. Clearly communicate how the technology will be used and ensure that users understand and agree to its implications. Ensure that the introduction of AR does not result in unfair treatment of workers. Be vigilant to prevent discrimination, whether intentional or unintentional, in the use of AR.

- *Transparency:* Be transparent about the decision-making processes involving AR, especially if automated decision-making algorithms are used. Clearly communicate how AR data is utilized and how it influences project decisions.

- *Environmental Impact:* Consider the broader impact of AR on the local community and the environment. Strive to minimize any negative consequences and communicate openly about the potential benefits and challenges of AR implementation.

10.8 CONCLUSION

Augmented reality has just broken onto the scene and has already been transforming construction. It is driving construction towards more technologically advanced, efficient, and client-focused outcomes. It benefits the entire construction team: engineers, designers, architects, project managers, and service providers. It will become a standard tool for conveying design changes for early adopters, eliminating the complexities of traditional design communication methods. It will facilitate visualization and decision-making based on real-time. Construction projects can have AR throughout the entire lifecycle. As a result, accuracy improves, and rework can be prevented, lowering building costs overall.

The integration of AR in the construction industry represents a transformative leap into a new era of efficiency, precision, and collaboration. Augmented reality is advancing at a rapid pace, and adoption across many industries is increasing. It is quickly becoming indispensable in the construction industry because it offers various benefits to contractors, engineers, architects, and other construction professionals. It is here to stay, and it is changing the game in construction. The future of AR in construction is promising and bright. More information about augmented reality in construction can be found in the books in [18-20] and a related magazine: *Construction Magazine*.

REFERENCE

[1] "7 Applications for augmented reality (AR) in construction,"

https://smarttek.solutions/blog/augmented-reality-in-construction/

[2] A. Miller, "Augmented reality and the construction industry," April 2009,

https://www.linkedin.com/pulse/augmented-reality-construction-industry-andrew-miller/

[3] M. N. O. Sadiku, C. M. M. Kotteti, and J. O. Sadiku, "Augmented reality in the construction industry," *International Journal of Trend in Research and Development*, vol. 12, no. 1, January-February 2024, pp. 71-78.

[4] M. N. O. Sadiku, C. M. M. Kotteti, and S. M. Musa, "Augmented reality: A primer," *International Journal of Trend in Research and Development*, vol. 7, no. 3, 2020.

[5] "What is augmented reality or AR?"

https://dynamics.microsoft.com/en-us/mixed-reality/guides/what-is-augmented-reality-ar/

[6] "Metaverse and multiverse: The real sense of AR, VR, MR, XR and IR,"

https://yp.ieee.org/metaverse-and-multiverse-the-real-sense-of-ar-vr-mr-xr-and-ir/

[7] "Introduction to augmented reality," June 2016,

https://www.informit.com/articles/article.aspx?p=2516729&seqNum=4

[8] H. Ling, "Augmented reality in reality," *IEEE MultiMedia*, July–September 2017, pp. 10-15.

[9] M. Singh and M. P. Singh, "Augmented reality interfaces," *IEEE Internet Computing*, November/December 2013, pp. 66-70.

[10] K. Arabi, "Explore the impact of augmented reality in construction (2024)," May 2023,

https://neuroject.com/augmented-reality-in-construction/

[11] D. Miller, "The potential of augmented reality in business," December 2018,

https://www.epikso.com/blog/the-potential-of-augmented-reality-in-business/

[12] N. Sahota, "Augmented reality in education: The future of learning is here," February 2023,

https://www.neilsahota.com/augmented-reality-in-education-the-future-of-learning-is-here/

[13] I. B. Alkan and H. B. Basaga, "Augmented reality technologies in construction project assembly phases," *Automation in Construction*, vol. 156, December 2023,

[14] A. Morozova, "How augmented reality is used in the construction industry to save money and speed up projects," November 2018,

https://medium.com/@info_35021/how-augmented-reality-is-used-in-the-construction-industry-to-save-money-and-speed-up-projects-c2696e03b7eb

[15] "How is augmented reality used in construction?" May 2023,

https://gamma-ar.com/how-is-augmented-reality-used-in-construction/

[16] "7 Uses in 2020 for augmented reality in construction," October 2019,

https://esub.com/blog/7-uses-in-2020-for-augmented-reality-in-construction/

[17] "Augmented reality in construction," May 2023,

https://arborxr.com/blog/augmented-reality-in-construction/

[18] A. Prabhakaram et al. (eds.), Applications of Immersive

Technology in Architecture, Engineering and Construction: A Handbook (Spon Research). Routledge, 2025.

[19] V. Sangiorgio et al., *New Approaches for Multi-Criteria Analysis in Building Constructions: User-Reporting and Augmented Reality to Support the Investigation*. Springer, 2022.

[20] M. Nourbakhsh, *Augment It: How Architecture, Engineering and Construction Leaders Leverage Data and Artificial Intelligence to Build a Sustainable Future*. Grammar Factory Publishing, 2022.

CHAPTER 11
BUILDING INFORMATION MODELING FOR CONSTRUCTION

"A great building must begin with the immeasurable, must go through measurable means when it is being designed, and in the end must be unmeasured."

—Louis Kahn

11.1 INTRODUCTION

The construction industry is undergoing fundamental changes. An innovative and time-saving tool called building information modeling (BIM) is responsible for that revolution. BIM is the foundation of digital transformation in the architecture, engineering, and construction (AEC) industry. It is a cutting-edge building design technology that starts with creating an intelligent 3D model. It is a digital representation of physical and functional characteristics of a building or other physical asset and facility. It rapidly enhances complex building processes, shortcutting any project completion date, cutting the expenses, and improving the overall quality of the construction. It allows construction managers and architectural professionals to communicate easily and build more efficiently, while giving clients a higher quality finished product. In short, it allows us to build a 3D model before the physical project is built. The fundamental use of BIM is to link people and technology with processes and workflows to enhance decision making and outcomes in architecture, engineering and construction (AEC).

The construction industry is considered by some to be behind other industries when it comes to adopting new technologies. AEC companies use a variety of software packages to design, commission, and operate assets. The construction industry is ever evolving with new and exciting technologies that arise everyday. Figure 11.1 shows a typical construction site [1], while Figure 11.2 shows some construction workers [2].

Figure 11.1 A typical construction site [1].

Figure 11.2 Some construction workers [2].

Building information modelling (BIM) is a holistic process most commonly used in the construction industry. It is one of the most promising developments in the architecture, engineering, and construction (AEC) industries. BIM is one of the disrupting technologies in the construction industry that allows the project stakeholders, primarily, architects, designers, engineers, and contractors to visualize how their contributions fit in the big picture. It is a digital representation of the physical and functional characteristics of a building or facility. This digital transformation is providing value across the construction industry. The main objective of BIM is to enhance project performance and produce better outcomes. BIM is transforming the way construction and infrastructure projects are planned, designed, constructed, and managed. It is a powerful driver leading the construction sector towards sustainability [3].

The aim of this chapter is to examine how building information modeling (BIM) is used in the construction industry. It begins with providing some background information on BIM. It describes BIM in construction. It highlights the benefits and challenges of BIM in the construction industry. It concludes with comments.

11.2 BACKGROUND ON BUILDING INFORMATION MODELING

Building information modeling (BIM) is a digital representation of physical and functional characteristics of a facility. It is the holistic process of creating and managing information for a built asset. BIM is a digital representation of the physical and functional characteristics of a building or facility. The representation can be used to simulate the construction and operation of the facility, and to provide information for decision-making throughout the project lifecycle [4,5].

Building information models (BIMs) are computer files (often but not always in proprietary formats and containing proprietary data) which can be extracted, exchanged or networked to support decision-making regarding a built asset. The main objective of creating the models is to provide builders with all relevant information about the project in a digital format. BIM software (such as Revit, Microstation, Vectorworks, Infraworks, ArchiCAD, BIM 360, Formit, Civil 3D, GBS Tekla, etc.) allows the design team on a project to construct a 3D model that represents the one we intend to build. Two typical examples of 3D models used in BIM are presented in Figure 11.3 [6] and Figure 11.4 [7].

Figure 11.3 A typical 3D model used in BIM [6].

Figure 11.4 Another example of 3D model used in BIM [7].

The BIMsoftware is used by individuals, businesses and government agencies who plan, design, construct, operate and maintain buildings and diverse physical infrastructures, such as water, refuse, electricity, gas, communication utilities, roads, railways, bridges, ports and tunnels. The BIM concept envisages virtual construction of a facility prior to its actual physical construction, in order to reduce uncertainty, improve safety, work out problems, and simulate and analyze potential impacts. However, the use of BIM goes beyond the planning and design phase of a project, extending throughout the life cycle of the asset. Figure 11.5 shows various components of BIM [8]. Figure 11.6 displays BIM project stakeholders, including architects, surveyors, engineers, contractors and the owner of the building itself [9].

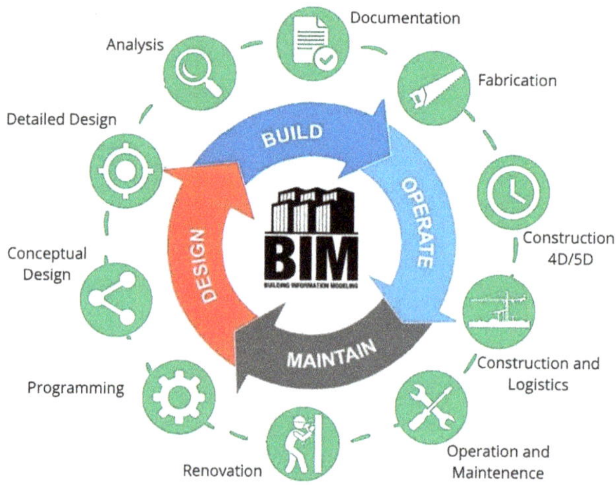

Figure 11.5 Various components of BIM [8].

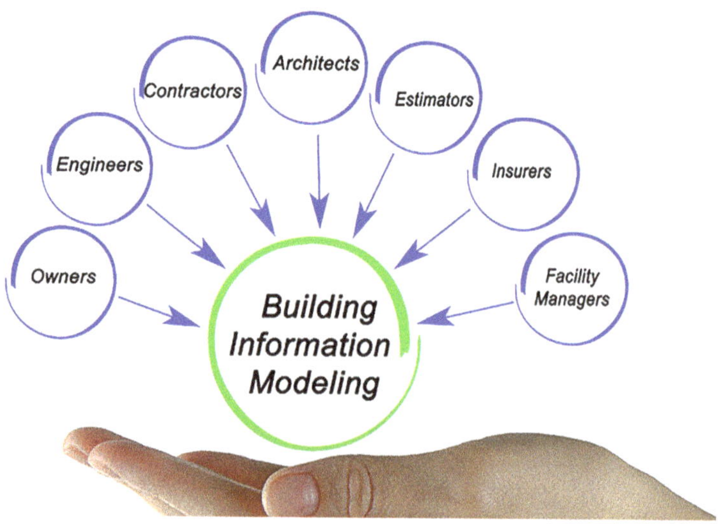

Figure 11.6 BIM project stakeholders [9].

Though the 'B' in BIM stands for building, it covers all constructions in the AEC industry, from residential and industrial buildings to bridges and roads. The 'I' in BIM stands for information, referring to the integrated information within BIM models. This is its core concept - information, parameters, attributes, whatever we call it. The 'M' in BIM stands for modelling. BIM models are known to be condensed with data. Thus, BIM is not a type of software, but a holistic process for managing the building information. We use software and tools that aid the BIM process. BIM covers the entire cycle of the building, from design development to post-construction maintenance. It includes planning, designing, construction, and even operations.

AEC professionals relied on manual drawings for decades until CAD came into existence. Traditional building design was largely reliant upon two-dimensional technical drawings (plans, elevations, sections, etc.). Building information modeling extends the three primary spatial dimensions (width, height and depth). The model can contain all the physical and functional characteristics of a project, including structure, geometry, aesthetics, materials, systems, and dynamic performance. A 3D model becomes 4D if a time component is added, and 5D when cost-related information is included. BIM enables a virtual information model to be shared by the design team (architects, landscape architects, surveyors, civil, structural and building services engineers, etc.), the main contractor and subcontractors, and the owner/operator. It enables all the stakeholders to use the same shared 3D model, from architects to surveyors, engineers, contractors, to building owners. This allows everyone to have access to the relevant information at the right moment during the design and construction of a project.

The concept of BIM has existed since the 1970s, when architects and engineers began to explore the use of computer-aided design (CAD) software to create 3D models of buildings. The terms "Building Information Model" or "Building Information Management," and "Building Information Modeling" did not become popularly used until some 10 years later. BIM only became an agreed term in the early 2000s. BIM has evolved over the past few decades from simple 2D drafting to sophisticated 3D modeling with rich data integration. Figure 11.7 shows early developments of BIM [10]. Today, BIM is an essential tool for the AEC industry. It is used throughout the entire project lifecycle, from design and construction to operation and maintenance.

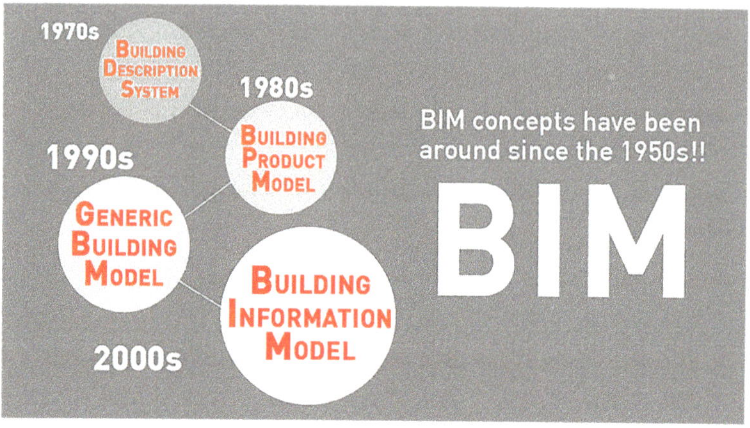

Figure 11.7 Early developments of BIM [10].

A comprehensive BIM modeling solution combines the major elements of building information modeling software, data repository, collaboration, analysis, visualization, and project management tools [7]. Different levels of BIM can be achieved for various types of projects. Below are brief descriptions of the levels [11]:

Level 0 BIM: Paper-based drawings + zero collaboration

Level 1 BIM: 2D construction drawings + some 3D modeling

Level 2 BIM: Teams work in their own 3D models

Level 3 BIM: Teams work with a shared 3D model

Level 4 BIM: Incorporating time-related data into the BIM model

Level 5 BIM: Integrating cost estimation and management within the BIM model

Level 6 BIM: Including sustainability data

Level 7 BIM: Used to manage and maintain facilities

11.3 BIM IN CONSTRUCTION

Beyond AEC, BIM is applied in other industries, including infrastructure, manufacturing, and facilities management, Other applications include running scenarios, such as the function of an emergency ventilation system, and energy analysis and modeling. Different applications of BIM are illustrated in Figure 11.8 [12]. Common areas of application of BIM include [8,12]:

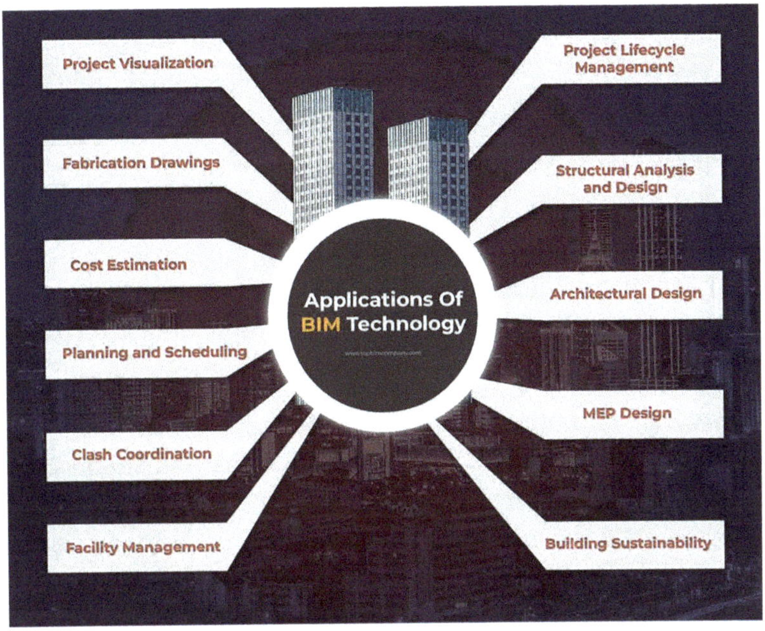

Figure 11.8 Different applications of BIM [12].

- *Architecture:* It is in the field of architecture that we might perhaps see the best uses of BIM. BIM application has allowed architects to participate in all phases of the project. Energy and building performance analyses can also be done which in turn produce more user and environment responsive designs.

- *Civil Engineering:* BIM technology in civil engineering enables structural analysis and design. BIM helps civil engineers digitally explore alternative design decisions, capture more detail, and enhance coordination. Using BIM increases accuracy, predictability, and understanding throughout project life cycles. BIM allows collating real capture data, CAD, and GIS data to create an optimized model fit for engineering works. The BIM models are rich in data and with the capacity to perform simulations and analysis. Advanced tools and plug-ins are available for road and bridge construction, reducing risk during and after construction while also lessening the negative environmental impact.

- *Architectural Design:* The architectural design process used to be a tedious and time-consuming process before the advent of BIM architecture. Figuring out various design iterations and diagrams for a building project was usually carried out using paper, pen, and CAD, which are not efficient tools. BIM helps in the architectural design of a building. This makes the architectural design process faster and more efficient.

- *Construction Management:* BIM implementation in construction management is an inevitable topic when discussing its uses in the AEC industry. Effective construction management is equivalent to effective handling of schedules, cost, safety, and quality through all stages of the project.

- *Facility Management:* BIM technology makes facility management smooth and streamlined. Facility Managers can use BIM throughout the facility's lifecycle to make smarter choices about things like space utilization, layout, equipment and asset upkeep, energy efficiency, and cost savings. From designing to 4D scheduling, 5D cost estimation, and facility management, BIM is an important ingredient across the project lifecycle.

- *Project Visualization:* BIM helps to visualize the project during the pre-construction stage, which enables the architects, engineers, and clients to find out the problems with the building, and fix them before construction. This helps to save materials, and costs considerably. BIM visualization also enables virtual reality and augmented reality.

- *Fabrication Drawings:* It is not difficult to produce shop drawings for the many different kinds of building systems with the help of BIM. The use of BIM for manufacturing allows the manufacturer to get precise details for fabrication. Fabrication drawings are automatically created while modeling the BIM model.

HVAC, electrical, plumbing, firefighting, façade, and structural shop drawings are some examples of shop drawings.

- *Cost Estimation:* 5D BIM helps in cost estimation. This feature of BIM technology is a boon in saving a considerable amount of money and enables the contractors to keep the building construction within the budget.

The application of BIM is not limited to these areas but is also used in other AEC fields such as urban planning, landscape, and structural engineering.

11.4 BENEFITS

As a result of comprehensive capabilities, BIM is becoming increasingly popular in the construction industry. Some benefits of BIM are displayed in Figure 11.9 [10].

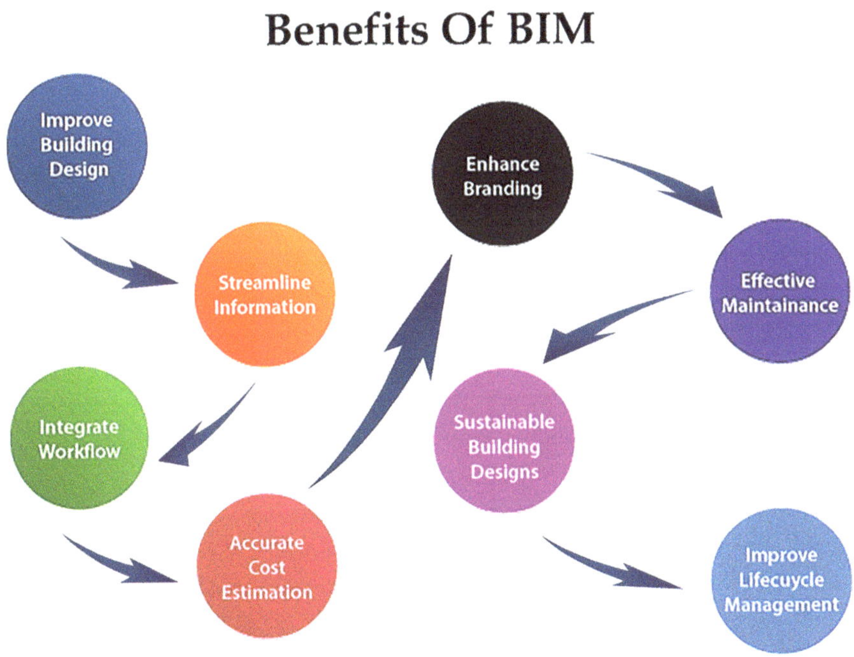

Figure 11.9 Some benefits of BIM [10].

BIM software allows all stakeholders in a construction project to collaborate in real-time. Using BIM gives you greater visibility, better decision-making, more sustainable options, and cost-savings on architecture, engineering, and construction (AEC) projects. Other benefits include [13]:

- *Collaboration:* At the heart of a BIM process is collaboration. Working as a collaborative team requires the development of better and more efficient ways of working to achieve shared goals. It brings about many advantages for the construction industry, such as improved communication and understanding, leading to greater productivity, quality, and cost certainty. Ultimately collaboration results in better outcomes. BIM enables collaboration across the project team –helping to optimize designs, improve accuracy, and connect design to fabrication to deliver projects faster and more efficiently.

- *Communication:* Communicating with stakeholders is simpler and can help speed up decision-making about the project. The BIM process helps all parties involved in a construction project to communicate easily. Everything is available in one place, and using cloud-based software means it is accessible from anywhere.

- *Centralization:* All aspects of the project are centralized. This reduces the chance of information loss that typically occurs when a new team takes ownership of a project. BIM serves as a centralized repository for all project-related information, including geometric data, specifications, materials, and performance characteristics. This information is accessible to all stakeholders involved in the project, ensuring transparency and consistency.

- *Global Usage:* BIM software is used by individuals, businesses, and government agencies worldwide. Across the world, BIM is a crucial and even mandated process to ensure the planning, design, and construction of buildings is highly efficient and collaborative. Most of the world's leading AEC firms are using BIM for nearly all of their projects. Many countries have adopted BIM standards and regulations to standardize processes and ensure interoperability. For example, the government of Hong Kong mandates the use of BIM for all government projects over HK$30M since 1 January 2018. Since 2010, the Korean government has been gradually increasing the scope of BIM-mandated projects. Figure 11.10

shows BIM mandate by country [14]. Mandates are simply government policies dictated for a requirement.

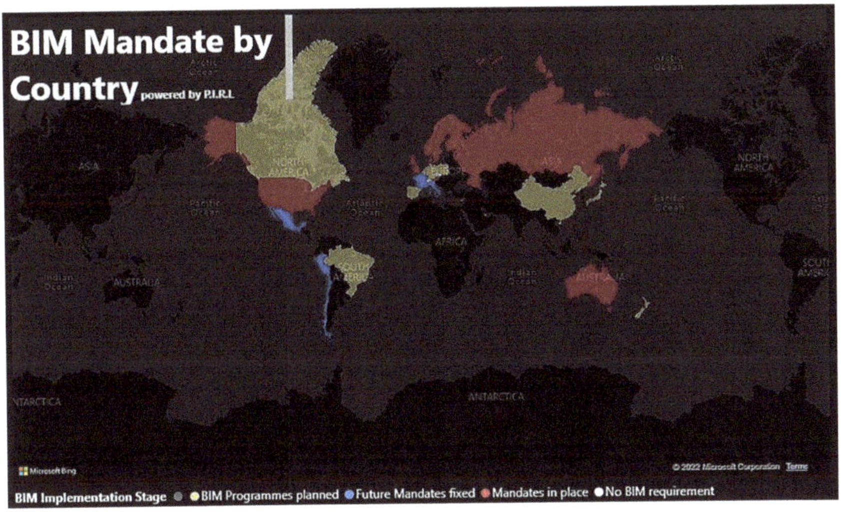

Figure 11.10 BIM mandate by country [14].

- *Green BIM:* BIM in green building, or green BIM, is a process that can help architecture, engineering, and construction firms to improve sustainability in the built environment. It can allow architects and engineers to integrate and analyze environmental issues in their design over the life cycle of the asset. Using BIM increases accuracy, predictability, and understanding throughout project life cycles. Through the time-efficiencies achieved with BIM, we can reduce energy use and cost. BIM helps reduce material waste during construction and building management, and it can eventually assist in sustainable demolition. Energy modeling using BIM can also minimize energy consumption over a building's life.

- *Error Elimination:* Because the virtual 3D building model is the source for all 2D and 3D drawings, design errors caused by inconsistent 2D drawings are eliminated. Conflicts and constructability problems are identified before they are detected in the field. Coordination among participating designers and contractors is enhanced and errors of omission are significantly reduced.

- *Construction Planning:* Construction planning using 4D CAD requires linking a construction plan to the 3D objects in a design, so that it is possible to simulate the construction process and show what the building and site would look like at any point in time. This graphic simulation provides considerable insight into how the building will be constructed day-by-day and reveals sources of potential problems and opportunities for possible improvements (site, crew and equipment, space conflicts, safety problems, and so forth).

- *Cost:* BIM helps in cost estimation and monitoring, leading to cost savings and budget adherence. BIM allows thorough cost analysis across a project. Reliable estimates for materials, shipping, and labor are made well before the construction stage begins. This helps project managers and teams reduce costs, by sourcing materials at their best possible price, selecting a more cost-effective material, or reducing any unnecessary labor.

- *Opportunity:* As BIM is helpful for predicting project outcomes, it can open a variety of opportunities. It is possible to view a 3D model using virtual reality which allows stakeholders, and other professionals to see the built environment before it exists. This provides more scope for innovation before investing in construction.

- *Efficiency:* BIM streamlines project scheduling and reduces construction time. It enables constant information exchange among architects, engineers, specialists, developers, contractors, and other parties. It helps teams to make well-informed decisions faster, thereby achieving more effective and efficient design than if BIM were not used.

- *Visualization:* BIM offers tools that help in proper planning and clear visualization before the initiation of the construction work. 3D visualization and the simulation of the surface area help the client to get the post-construction visualization of the infrastructure thus facilitating easy modifications prior to the construction stage.

- *Interoperability:* BIM software supports interoperability, allowing data exchange between different platforms and disciplines. This facilitates collaboration among architects, engineers, contractors, and other stakeholders, enabling seamless integration of design and construction processes.

- *Reduced Waste:* In today's construction environment, labor is in short supply, timelines are compressed, and budgets are tighter. Eliminating waste becomes a critical goal. By using building information modeling, HVAC materials estimates can be exact and fabrication waste is reduced. By optimizing on-site labor through efficient design, BIM helps reduce waste from fabrication.

- *Clash Detection:* BIM software detects clashes and discrepancies in the design phase, preventing expensive revisions during the construction process. This guarantees a smooth workflow and mitigates potential project setbacks.

11.5 CHALLENGES

Despite the demonstrated benefits from numerous case studies, there still remains a few BIM challenges that continue to hinder the wider adoption and implementation of BIM. There is no clear consensus on how to implement or use BIM. There is no single BIM document providing instruction on its application and use. There is also a need to outline who should develop and operate the building information models and how the developmental and operational costs should be distributed. Other challenges include [10,14]:

- *Increased Adoption:* BIM adoption is likely to continue growing worldwide. Governments and industry bodies in various countries have already mandated or encouraged the use of BIM in construction projects, and this trend is expected to continue.

- *Interoperability:* Improving interoperability between different BIM software and data formats is a significant challenge that the industry will continue to address. Open standards and collaborative efforts will be crucial in achieving better data exchange and integration.

- *Integration with Emerging Technologies:* BIM will integrate with other emerging technologies like artificial intelligence (AI), machine learning (ML), augmented reality (AR), virtual reality (VR), and mixed reality (MR). These technologies can help in visualizing and interacting with BIM models in more immersive and interactive ways. Artificial intelligence and machine learning will be used to analyze and extract insights from BIM data, optimizing designs, identifying potential issues, and aiding in decision-making throughout the construction process.

- *Generative Design:* BIM software is expected to incorporate generative design tools that can automatically create and evaluate design options based on specific project parameters, leading to more innovative and efficient designs.
- *Regulation and Standards:* Regulations and standards related to BIM will continue to evolve as the technology matures and becomes more prevalent. Staying up-to-date with these changes will be essential for industry professionals.
- *No Standardized Mandates:* Mandates are simply government policies dictated for a requirement. BIM mandates require its implementation in projects. To this day, only a few countries have BIM mandates, and even then they are not strict or required for all project types. Furthermore, most, if not all, are developed nations where training for BIM is more readily available.
- *Education:* BIM has not been extensively taught as part of the curriculum, especially during undergraduate studies. Only the basics are covered. Undergraduate students may be familiar with software like Revit and may have used it for their coursework. As BIM becomes increasingly important in the AEC industry, there will be a greater emphasis on educating and training professionals in BIM skills and best practices.
- *Lack of Trained Professionals:* Many construction companies have complained there is a lack of skilled professionals. The aftermath is an inefficient workflow with errors and profit

loss. Such BIM challenges can make the firms even more hesitant to train existing staff or hire those with skills due to the misconception that BIM technology is difficult and inefficient to adopt.

- *Cost:* BIM software can be expensive. As the need to train the existing employees or hire BIM educated professionals overwhelms, BIM implementation becomes seen as an expense rather than an investment. Most of these firms would not even realize it would have a high ROI (return on investment).

- *Change:* There is hesitation and unwillingness of stakeholders to adopt BIM. Another reason for the hesitation is that along with the implementation comes the necessary change in operation. Therefore, rather than investing in training or hiring, the firms stick to the more traditional methods. The gap between the skilled and the unskilled becomes noticeable, adding to the reluctance.

11.6 CONCLUSION

BIM technology is a revolutionary approach to the design, construction, and management of buildings and infrastructure. It is one of the most promising developments in the architecture, engineering, and construction (AEC) industries. The construction industry is yet to realize the full capabilities of BIM. Implementing BIM requires a shift in the way stakeholders in the construction industry work together. It requires collaboration, communication, and a willingness to share information.

BIM is a game-changer for the AEC industry. It has become the standard for modern construction projects. In the coming years, BIM adoption rates are likely to increase. Pressure on the industry is intensifying as the population grows, and BIM processes help meet these demands. The global growth of BIM is indisputable and we can only expect it to soar as more and more companies and countries are adopting BIM. More information on BIM for construction is available from the books in [15-28] and in the following related journals:

- *Construction Magazine*
- *Buildings*
- *Journal of Building Engineering*
- *Journal of Construction Engineering and Management*
- *WIT Transactions on The Built Environment*

REFERENCE

[1] "What is BIM and how does it impact the construction industry?"

https://www.faro.com/en/Resource-Library/Article/BIM-and-how-does-it-impact-the-construction-industry

[2] "What is building information modeling? (BIM)," January 2021,

https://www.accruent.com/resources/blog-posts/what-building-information-modeling

[3] M. N. O. Sadiku, P. A. Adekunte, and J. O. Sadiku, "Building information modeling for construction," *International Journal of Trend in Scientific Research and Development*, vol. 9, no. 1, January-February 2025, pp. 142-152.

[4] "BIM (building information modelling)" Unknown Source.

[5] "What is building information modelling (BIM)?" Unknown Source.

[6] "Building information modeling," *Wikipedia*, the free encyclopedia,

https://en.wikipedia.org/wiki/Building_information_modeling

[7] O. Harris, "What are the key components of BIM modeling solution?"

https://vocal.media/education/what-are-the-key-components-of-bim-modeling-solution

[8] T. Hnin, "Understanding what is BIM: The complete guide 2025," May 2022,

https://www.novatr.com/blog/detailed-bim-guide

[9] "Top 5 benefits of BIM for HVAC,"

https://www.tcgduct.com/top-5-benefits-of-bim-for-hvac/

[10] M. Sharon, "Complete guide to building information modeling (BIM)," November 2023,

https://medium.com/@matt-sharon/complete-guide-to-building-information-modeling-bim-ea0dd2cf922e

[11] S. Lorek, "What is BIM (building information modeling)," April 2022,

https://constructible.trimble.com/construction-industry/what-is-bim-building-information-modeling

[12] "Applications of BIM technology in the construction industry,"

https://www.topbimcompany.com/applications-of-bim-technology/#:~:text=Visualization%2C%20fabrication%2C%20coordination%2C%20cost,the%20BIM%20applications%20in%20construction.

[13] A. Patel, "BIM definition, software tools & benefits," September 2022,

https://www.united-bim.com/building-information-modeling/

[14] T. H. S. Aung, "The 6 challenges of implementing BIM faced by the AEC industry,"

https://www.novatr.com/blog/challenges-of-implementing-bim

[15] W. Kymmell, *Building Information Modeling: Planning and Managing Construction Projects with 4D CAD and Simulations (McGraw-Hill Construction Series): Planning and ... Projects with 4D CAD and Simulations*. McGraw Hill 2008.

[16] M. De Las Nieves et al., *New Advances In Building Information Modeling And Engineering Management: Digital Innovations In Architecture, Engineering And Construction*. Springer Cham, 2023.

[17] J. U. Msed and U. Isikdag (eds.), *Handbook of Research on Building Information Modeling and Construction Informatics: Concepts and Technologies*. Information Science Reference, 2009.

[18] R. Crotty, *The Impact of Building Information Modelling:*

Transforming. Routledge, 2016

[19] S. Mordue, P. Swaddle, and D. Philp, *Building Information Modeling For Dummies*. For Dummies, 2015.

[20] O. Koseoglu and Y. Arayici, *Airport Building Information Modelling. Routledge*, 2019.

[21] R. Garber, *BIM Design: Realising the Creative Potential of Building Information Modelling (AD Smart)*. Wiley, 2014.

[22] Y. Arayici et al. (eds.), *Heritage Building Information Modelling. Routledge*, 2017.

[23] A. H. Baik, *Heritage Building Information Modelling for Implementing UNESCO Procedures: Challenges, Potentialities, and Issues*. Routledge, 2020.

[24] P. Wu, H. Li, and X. Wang (eds.), *Integrated Building Information Modelling*. Bentham Science Publishers, 2017.

[25] K. Kensek and D. Noble, *Building Information Modeling: BIM in Current and Future Practice*. Wiley, 2014.

[26] G. Hofmeister, Building Information Modelling: Overview Of The Bim For The Beginners And Serves As Pathfinder. Independently Published, 2022.

[27] M. Dastbaz, C. Garse, and A. Moncaster (eds.), *Building Information Modelling, Building Performance, Design and Smart Construction*. Springer, 2017.

[28] S. Race, *BIM Demystified: An Architect's Guide to Building Information Modelling/Management (BIM)*. RIBA Publishing, 2nd edition, 2013

CHAPTER 12
CYBERSECURITY IN CONSTRUCTION

"As construction becomes more digitally integrated, robust cybersecurity practices are essential to safeguard project timelines, finances, and critical infrastructure."

—Anonymous

12.1 INTRODUCTION

The construction industry, a cornerstone of global infrastructure development, is undergoing a digital transformation, but this progress comes with increased cybersecurity vulnerabilities. The industry is now confronting a new frontier of threats emanating from the digital realm. Today, the built environment is designed, constructed, and managed using digital technology, making it increasingly exposed to cybersecurity risks. The construction industry grapples with ensuring the security of its physical assets and its digital infrastructure. As a result, the industry has experienced massive losses including stolen or misdirected funds.

From building neighborhood homes to mile-long suspension bridges, construction is the foundation of the growth and essential to the development and advancement of any nation. The construction industry is at a crossroads of innovation and vulnerability. Although digital innovations can help ensure that projects remain under budget on construction sites, the growing digitalization may present additional challenges. Although smartphones, laptops, and tablets are widely used in the construction industry to monitor data, transfer information, and send communications to employees and clients, it can be challenging to enact widespread security measures with so many devices. Projects often involve numerous partners, increasing the complexity of managing cybersecurity across different systems. This complexity presents numerous entry points for cybercriminals seeking to exploit vulnerabilities within interconnected systems. Figure 12.1 shows a typical construction site [1], while Figure 12.2 shows some construction workers [1].

Figure 12.1 A typical construction site [1].

Figure 12.2 Construction workers [1].

The construction industry faces several cybersecurity threats due to its unique nature. The industry is increasingly becoming a prime target for cyberattacks. Data breaches, supply chain attacks, and ransomware pose significant threats. For years, cybersecurity experts have warned construction industry and workers that they are targets for ransomware attacks, phishing theft, data breaches, and theft of sensitive information. Cyber criminals see the construction sector as a potential weak and easy target to attack. By prioritizing cybersecurity, construction companies can protect their financial health, maintain business continuity, and preserve their hard-earned reputations [2].

This chapter discusses the role of cybersecurity in the construction industry. It begins with explaining what cybersecurity is all about. It describes cybersecurity in construction and provides various ways to combat it. It highlights the benefits and challenges of cybersecurity in the construction industry. It concludes with comments.

12.2 OVERVIEW ON CYBERSECURITY

Cybersecurity refers to a set of technologies and practices designed to protect networks and information from damage or unauthorized access. It is vital because governments, companies, and military organizations collect, process, and store a lot of data. As shown in Figure 12.3, cybersecurity involves multiple issues related to people, process, and technology [3]. Figure 12.4 shows different components of cybersecurity [4].

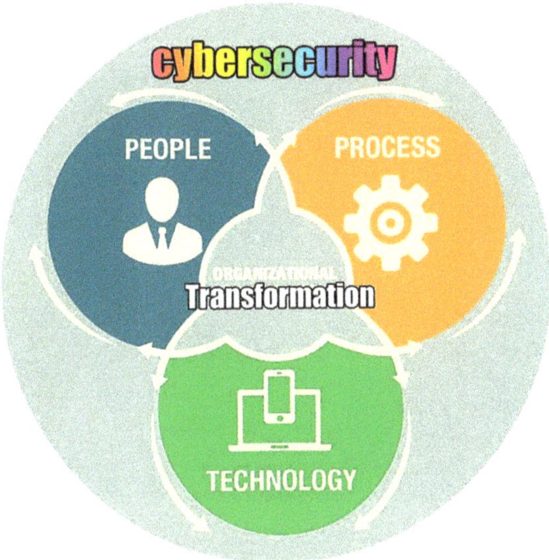

Figure 12.3 Cybersecurity involves multiple issues related to people, process, and technology [3].

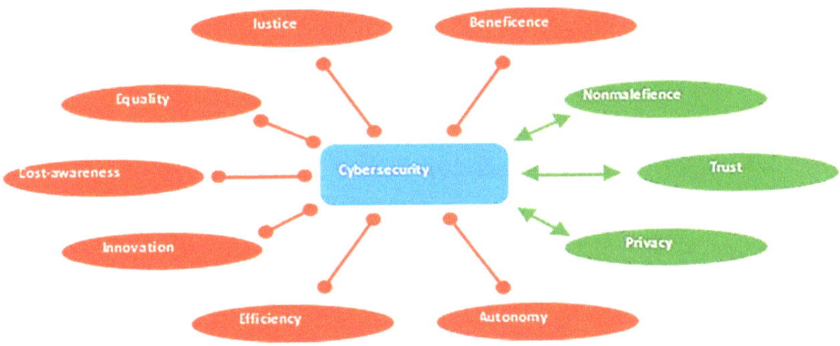

Figure 12.4 Different components of cybersecurity [4]. (Green: supportive; Red: in tension).

A typical cyberattack is an attempt by adversaries or cybercriminals to gain access to and modify their target's computer system or network. Cybercriminals or ethical hackers are modern-day digital warriors, possessing extraordinary skills and knowledge to breach even the most impregnable systems. A typical cybercriminal is shown on Figure 12.5 [5]. Cyber attacks are becoming more frequent, sophisticated, dangerous, and destructive. They are threatening the operation of businesses, banks, companies, and government networks. They vary from illegal crime of individual citizen (hacking) to actions of groups (terrorists) [6].

Figure 12.5 A typical cybercriminal [5].

The cybersecurity is a dynamic, interdisciplinary field involving information systems, computer science, and criminology. The security objectives have been availability, authentication, confidentiality, nonrepudiation, and integrity. A security incident is an act that threatens the confidentiality, integrity, or availability of information assets and systems [7]. These are known as the pillars of information assurance.

- *Availability:* This refers to availability of information and ensuring that authorized parties can access the information when needed. Attacks targeting availability of service generally leads to denial of service.

- *Authenticity:* This ensures that the identity of an individual user or system is the identity claimed. This usually involves using username and password to validate the identity of the user. It may also take the form of what you have such as a driver's license, an RSA token, or a smart card.

- *Integrity:* Data integrity means information is authentic and complete. This assures that data, devices, and processes are free from tampering. Data should be free from injection, deletion, or corruption. When integrity is targeted, nonrepudiation is also affected.

- *Confidentiality:* Confidentiality ensures that measures are taken to prevent sensitive information from reaching the wrong persons. Data secrecy is important especially for privacy-sensitive data such as user personal information and meter readings.
- *Nonrepudiation:* This is an assurance of the responsibility to an action. The source should not be able to deny having sent a message, while the destination should not deny having received it. This security objective is essential for accountability and liability.

Good practices for cybersecurity in construction companies should include all of these elements.

Everybody is at risk for a cyber attack. Cyber attacks vary from illegal crime of individual citizen (hacking) to actions of groups (terrorists). The following are typical examples of cyber attacks or threats [8]:

- *Malware:* This is a malicious software or code that includes traditional computer viruses, computer worms, and Trojan horse programs. Malware can infiltrate your network through the Internet, downloads, attachments, email, social media, and other platforms. Spyware is a type of malware that collects information without the victim's knowledge.
- *Phishing:* Criminals trick victims into handing over their personal information such as online passwords, social security number, and credit card numbers.
- *Denial-of-Service Attacks:* These are designed to make a network resource unavailable to its intended users. These can prevent the user from accessing email, websites, online accounts or other services.
- *Social Engineering Attacks:* A cyber criminal attempts to trick users to disclose sensitive information. A social engineer aims to convince a user through impersonation to disclose secrets such as passwords, card numbers, or social security number.

- *Man-In-the-Middle Attack:* This is a cyber attack where a malicious attacker secretly inserts him/herself into a conversation between two parties who believe they are directly communicating with each other. A common example of man-in-the-middle attacks is eavesdropping. The goal of such an attack is to steal personal information.

These and other cyber attacks or threats are shown in Figure 12.6 [9]. Sources of cyber threats are displayed in Figure 12.7 [10].

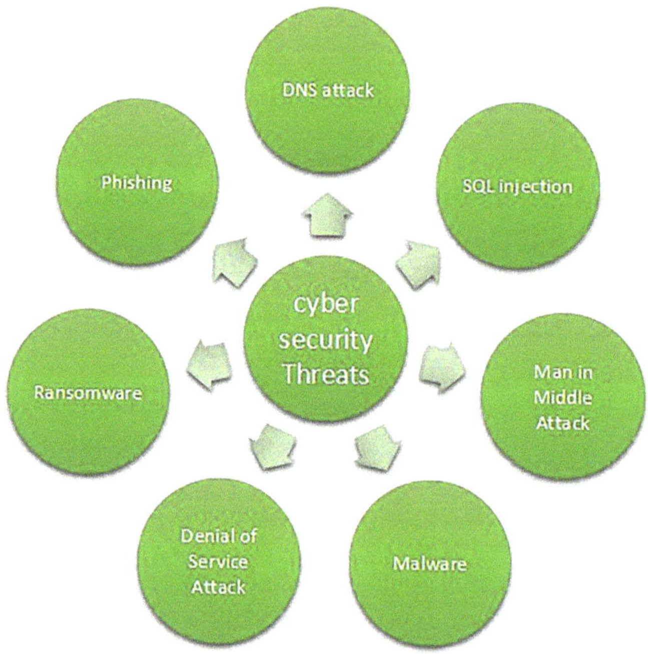

Figure 12.6 Common types of cybersecurity threats [9].

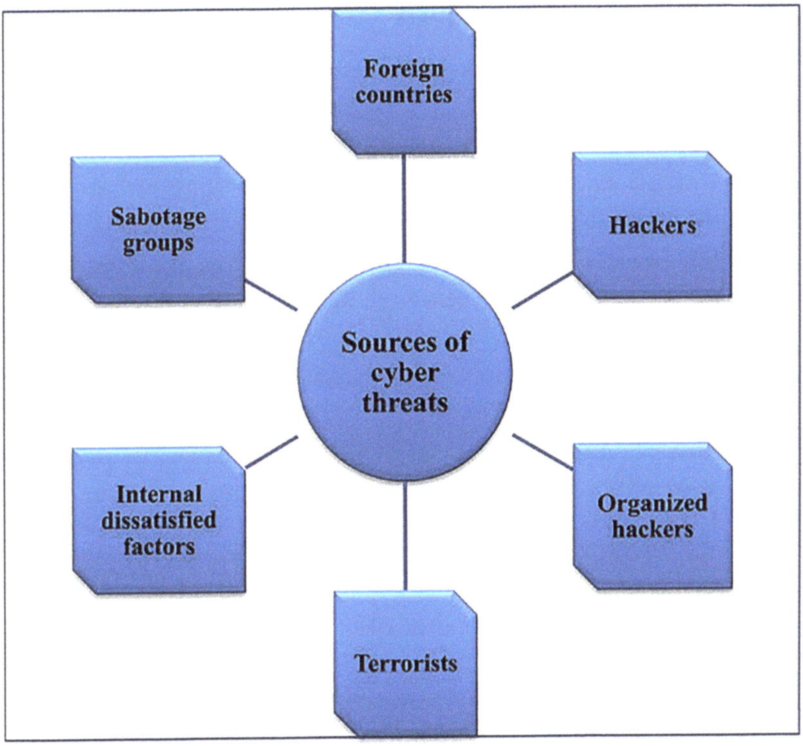

Figure 12.7 Sources of cyber threats [10].

The social and financial importance of cybersecurity is increasingly being recognized by businesses, organizations, and governments. Cybersecurity involves reducing the risk of cyber attacks. Cyber risks should be managed proactively by the management. Cybersecurity technologies such as firewalls are widely available [11]. Cybersecurity is the joint responsibility of all relevant stakeholders including government, business, infrastructure owners, and users. Cybersecurity experts have shown that passwords are highly vulnerable to cyber threats, compromising personal data, credit card records, and even social security numbers. Governments and international organizations play a key role in cybersecurity issues. Securing the cyberspace is of high priority to the US Department of Homeland Security (DHS). Vendors that offer mobile security solutions include Zimperium, MobileIron Skycure, Lookout, and Wandera.

12.3 CYBERSECURITY IN CONSTRUCTION

Cybersecurity attacks are making headlines these days, and the construction industry is not immune. In reality, all types of businesses are a target for cyberattacks. The way the construction industry operates makes construction companies especially vulnerable to attack. Integrating digital technologies has also put the industry at a higher risk for damages caused by cyberattacks. From data breaches compromising sensitive project information to sophisticated ransomware attacks disrupting operations, the construction industry finds itself at the intersection of technological innovation and cybersecurity vulnerability. Construction cybersecurity is a crucial part of business for all companies in the industry and will continue to be vital in the coming years.

As shown in Figure 12.8, cybersecurity in an important consideration in construction [12]. Cybercriminals love to target the construction industry for the following reasons [13]:

Figure 12.8 Cybersecurity in an important consideration in construction [12].

- The frequent transfer of sensitive information, such as blueprints, client, financial, and personal information, which can all be stolen or used as ransom.
- The unique structure of construction, with many remote workers (on site) and office locations, constantly communicating through digital networks and mobile devices such as emails and texts.
- Being an industry with a historically low perception of

risk regarding cyber threats. Many contractors or workers underestimate how big of a target they are, leading to underinvestment in protective measures.

- The complexity of the construction industry, involving many different possible points of failure, such as subcontractors, suppliers, clients, and workers.

There are several aways cyber-attacks could impact a construction company: (i) ransomware; (ii) fraudulent wire transfer; (iii) downtime or business interruption; (iv) breach of intellectual property; and (v) breach of bid data. Here are common cybersecurity threats in the construction industry [14,15]:

• *Data Breaches:* Construction companies handle vast amounts of sensitive data, including financial information, intellectual property, project designs, and client details. Data breaches can occur through hacking, malware, or social engineering attacks, leading to significant financial losses and reputation damage.

• *Supply Chain Attacks:* Construction projects involve diverse players, including architects, engineers, subcontractors, and suppliers, each contributing to the intricate tapestry of a project's lifecycle. Each entity represents a potential entry point for cybercriminals to infiltrate the network, compromise sensitive information, or disrupt operations.

• *Internet of Things (IoT) Vulnerabilities:* The increasing adoption of IoT devices in construction, such as connected sensors, drones, and wearable technology, introduces new cybersecurity risks. These devices often lack robust security measures and can be exploited by attackers. Construction companies increasingly focus on securing IoT devices by implementing network segmentation,

• *Ransomware:* Ransomware, or when a threat actor holds a computer system hostage for payment, can limit a construction company's access to critical systems and potential delay work at Project. Ransomware attacks have become a significant concern for the construction industry. Cybercriminals use malicious software to encrypt critical files and demand payment for their release, disrupting project timelines and causing financial harm. The most common delivery source for ransomware to infect an operating system is via phishing emails that seek to induce the reader to download malware.

- *Fraudulent Wire Transfers:* The number of repeated payments throughout a project involving many trade contractors, the prevalent use of apps to automatically process monthly payments, and the extensive use of tablets on job sites make the industry vulnerable to cybersecurity incidents. As construction companies become increasingly reliant on online banking and wire transfers, they become susceptible to wire fraud. Fraudulent wire transfers, often the result of social engineering, present a substantial risk the construction industry, who are often moving large sums of capital around. Falling victim to fraudulent wire transfer not only presents dire fiscal issues for a construction company but can also lead to server reputation harm.

- *Physical Security Risks:* Construction sites are often physically exposed and vulnerable to theft, vandalism, and unauthorized access. Cyber-physical attacks targeting equipment or building systems, such as HVAC or access control systems, can disrupt operations and compromise safety.

- *Business Interruption:* Operational disruptions are another significant effect of cyber incidents. The construction industry is heavily reliant on the ability to deliver projects on a deadline. A cyber-attack on a construction company's software or equipment could potentially cause a delay in the project while the cyber-attack is properly addressed.

- *Breach of Intellectual Property:* If a construction company is holding highly sensitive blueprints or schematics in its computer system, breach of these computer systems could result in major reputational damage and potential lawsuits.

- *Breach of Bid Data:* Any breach can lead to financial loss, legal repercussions, and a damaged reputation. If a construction company holds information regarding its bidding strategies on a computer system, access and acquisition of these files could lead to a loss of a competitive edge.

- *Data Theft:* If data is lost or stolen, operations can be severely disrupted, leading to project delays. The construction industry faces frequent attacks to steal intellectual property and private data. Cyber criminals commit data theft because social security and credit card numbers, as well as personal information of employees, vendors, and customers are very valuable to them. Whether acquired through ransomware or other hacking mechanisms, data theft is a serious issue. Some construction entities deal with sensitive and confidential intellectual property such as blue prints, designs, patents and bid information. Sophisticated hackers recognize the value of such information, and there is no limit to what they will do to extort their victims.

12.4 COMBATING CYBERSECURITY IN CONSTRUCTION

Escalating cyber threats, tightening regulations, and increasing reliance on technology has created a need for rapid cyber maturity and tailored risk transfer solutions so that construction firms can better weather these challenges. To combat these threats, construction professionals are implementing various cybersecurity measures:

- *Proactive Steps to Boost Cybersecurity:* A proactive approach to cyber security risks is essential. Construction companies must recognize the growing cyber threats they face and take proactive steps to safeguard their operations. Just as cybercriminals develop increasingly sophisticated hacking methods, cybersecurity companies continue to evolve the cybersecurity measures and systems available for company use. Partner with cybersecurity professionals who understand the unique challenges of the construction industry. They can provide tailored solutions, conduct penetration testing, and offer ongoing support to enhance your security posture. Have strong plans in place for the "what if" scenarios and ensure everyone is trained in the risk/response strategies and their immediate implementation.

- *Implement an Incident Response Plan:* Determine actions to be taken after a potential cyberattack to mitigate losses. Create an incident response plan, a plan of action for what your company will do if it falls victim to a cyber-attack. You need to be ready to take immediate action to mitigate damages. Determine which individuals in your company will be responsible for taking key actions, and have a point person designated to lead the charge.

- *Secure Supply Chain Management:* Work closely with subcontractors and suppliers to ensure they adhere to robust cybersecurity standards. Incorporate cybersecurity clauses in contracts and conduct regular audits to verify compliance. Using foresight can help mitigate the possible cybersecurity risks of working with subcontractors and suppliers.

- *Choosing the Right Software:* Adding robust cybersecurity measures to cover all devices and users is vital, but choosing the right software can help to shore up security from within.

- *Education:* Educate all of your team members about cybersecurity risks. Let them understand the cyber risks affecting your industry and how to identify and report social engineering attempts. Make sure that your personnel understand this risk and other vulnerabilities by conducting periodic training and workshops. It is likely not enough to just train your employees. You should periodically evaluate whether the lessons are sticking by sending mock phishing emails to see how well your employees perform. Make sure that your employees know who the point person is at your company to contact in the event of a cybersecurity breach.

- *Insurance:* Even with a robust cybersecurity defense, no system is immune to attacks. Consider having cyber liability insurance in place to help recover from worst case scenario outcomes. Cyber insurance helps mitigate exposure to cyber-related losses, filling gaps that may be left over from other policies (e.g., commercial property insurance, general liability insurance, etc.). If you suffer a data breach, your coverage may help you cover the costs of hiring attorneys, forensic IT consultants, and other crisis management costs that you may incur. Insurance may cover losses from fraudulent wire transfers and may also assist in negotiating and paying a ransom.

12.5 BENEFITS

A successful cyberattack can lead to damaging intellectual property theft, sensitive data breaches, ransomware, spyware, malware, disruption, and supply chain disorder. Government agencies and industry organizations play a crucial role in enhancing cybersecurity across the construction sector. By integrating cybersecurity into every phase of the project lifecycle, a company can protect sensitive data and maintain operational integrity. Other benefits include [16]:

- *Network Security:* Construction companies are deploying robust network security solutions, such as firewalls, intrusion detection systems, and encryption protocols, to safeguard their digital infrastructure from unauthorized access and malware attacks.

- *Supply Chain Management:* Implementing stringent cybersecurity requirements for vendors and subcontractors can help mitigate the risk of supply chain attacks. Contracts should include clauses addressing data protection and cybersecurity standards.

- *Secure IoT Deployment:* Construction firms increasingly focus on securing IoT devices by implementing network segmentation, regularly updating firmware, and monitoring device activity for signs of compromise.

- *Physical Security Measures:* Physical security measures such as surveillance cameras, access controls, and perimeter fencing can help deter unauthorized access to construction sites and protect valuable equipment and materials.

- *Regulatory Compliance:* Adhering to industry-specific cybersecurity regulations and standards, such as the NIST Cybersecurity Framework or GDPR, can help construction companies establish comprehensive cybersecurity policies and practices.

- *Ensuring Project Continuity:* Cyberattacks can disrupt project timelines, leading to delays and financial losses. By implementing robust cybersecurity protocols, construction companies can ensure the continuity of their projects and maintain the trust of clients and stakeholders.

- *Safeguarding Intellectual Property:* Architectural designs, proprietary construction methodologies, and sensitive project data are intellectual property treasures. Cybersecurity measures are essential to prevent the theft or compromise of these assets, preserving the competitive edge of construction firms.

- *Adherence to Regulatory Standards:* As the digital landscape evolves, so do regulatory standards. Cybersecurity is not only a defense against attacks but also a means of staying compliant with industry-specific regulations, avoiding legal repercussions.

12.6 CHALLENGES

Apart from handling a lot of big data, to operate, the construction industry relies on remote working, and often, a BYOD (bring your own device) culture, both of which are ways of working that create extra vulnerabilities within an IT infrastructure. It is urgent for construction companies to act to protect themselves from the devastating attacks which threaten to jeopardize project timelines, financial stability, reputations, and sensitive data. Other challenges include [16]:

- *Lack of Cybersecurity Infrastructure:* An entity without appropriate cyber hygiene and cyber architecture signifies an entity that is easy to attack and extort. Monetarily driven cyber criminals will be able to apply little effort for maximum gain. Many construction companies have not properly invested in cybersecurity and pay dearly when they experience an attack.

- *Target for Sensitive Information:* Cybercriminals may be able to access confidential data and then use this sensitive information to enact ransomware attacks. For nation states seeking to gain valuable infrastructure information, intellectual property, or entrance to critical public works, the construction industry is the weak link and an easy target for access. The threat actors seek to extort money, and the construction industry presents a big, lucrative target.

• *Integration of New Technologies:* Engineering and construction services have supported technologies such as artificial intelligence, building information modelling (BIM), machine learning, IoT, and robotics. The integration of new technologies makes processes so much easier and more efficient, yet it also introduces new challenges, making cybersecurity more essential than ever. Data privacy risk is often overlooked in the race to embrace new technologies, creating a significant risk.

• *Legacy Systems:* Many construction companies still rely on outdated software and hardware, which are more susceptible to vulnerabilities. Integrating these legacy systems with newer technologies without proper security measures can open the door to cyber threats. Support on legacy systems is a significant problem in the construction industry. Legacy or end-of-life operating systems present significant opportunities for cybercriminals.

• *Third-party Risk:* Construction companies often rely on multiple subcontractors and third-party vendors, which increases the potential for cybercriminals to target less secure partners. Once a third-party vendor's system is breached, attackers can gain entry into the main company's network, compromising sensitive data. Third-party cyber risk includes potential data breaches due to vulnerabilities within a vendor's IT environment and can lead to financial, reputational, and regulatory/compliance consequences.

• *Lack of Cybersecurity Regulations:* The construction industry has historically not been subject to mandatory cyber regulatory requirements or scrutiny. For many decades, it seemed the construction sector did not have many regulations in place for data security, whereas sectors like financial services are subject to stringent regulation. To date the construction industry has avoided the strict data privacy and security regulations that industries such as healthcare and banking have faced. The limited regulation and guidance in the construction industry may have contributed to less focus on cybersecurity than in other industries. However, the

US government has been increasingly regulating and requiring government contractors to comply with the National Institute of Standards and Technology (NIST) Cybersecurity Framework.

- *Data Breach Expenses:* When a threat actor accesses or acquires Personal Identifiable Information as defined by applicable law, your company has suffered a data security incident. Cyber liability insurance policies typically cover the costs of hiring of lawyers, forensic IT security vendors, public relations, or crisis communication costs to assist you in handling your response.

- *Cybersecurity Training:* Human error accounts for a high proportion of cybersecurity infractions. All employees, subcontractors, and temporary workers should be thoroughly trained in cyberattack prevention essentials. Training employees on good security practices —such as regularly installing software updates, creating strong passwords, utilizing multi-factor authentication on all devices, and encrypting sensitive data, can prevent slip-ups that might have severe effects.

- *Reputational Damage:* Trust is paramount in the construction industry. A breach can erode client confidence, leading to loss of future business opportunities and long-term harm to the company's reputation. The reputational damage that often comes with data breaches further burdens construction companies. Also, any blueprints, designs, methodology, patents, or other proprietary intellectual property is at serious risk if appropriate steps to mitigate cyberattacks are not taken.

- *Protection of Digital Assets:* With the industry's increased reliance on building information modelling (BIM), IoT devices, and cloud-based solutions, the protection of digital assets has never been more critical. Cybersecurity measures serve as a shield against unauthorized access and data breaches.

12.7 CONCLUSION

The construction industry is confronted with various cybersecurity challenges as it continues to adopt digital technologies and innovative solutions. The industry's digital evolution requires a proactive and comprehensive approach to cybersecurity. Cybersecurity should be of the utmost importance for construction firms today, but the industry should exercise caution. Embracing cybersecurity is no longer an option, but a necessity. By investing in cybersecurity, construction companies can build not just structures, but also resilience.

The construction industry faces an increasing threat from cyberattacks. Cybersecurity is becoming more essential as the danger and potential risks continue to rise and evolve. It should be fundamental to a strong corporate governance strategy that supports building trust among stakeholders. As the digital transformation accelerates, construction companies must recognize that cybersecurity is critical to their bottom line and business continuity.

REFERENCE

[1] "Why is cyber security important in the construction industry?" August 2024,

https://tinyurl.com/3tyyyhj2

[2] M. N. O. Sadiku, P. A. Adekunte, and J. O. Sadiku, "Cybersecurity in construction industry," *International Journal of Trend in Scientific Research and Development*, vol. 9, no. 1, January-February 2025, pp.102-111.

[3] P. Singh, "A layered approach to cybersecurity: People, processes, and technology- explored & explained," July 2021,

https://www.linkedin.com/pulse/layered-approach-cybersecurity-people-processes-singh-casp-cisc-ces

[4] M. Loi et al., "Cybersecurity in health – disentangling value tensions," *Journal of Information, Communication and Ethics in Society*, June 2019,

https://www.emerald.com/insight/content/doi/10.1108/JICES-12-2018-0095/full/html

[5] M. Adams, "Unlocking the benefits of ethical hacking: The importance of ethical hackers in cybersecurity," April 2023,

https://www.businesstechweekly.com/cybersecurity/network-security/ethical-hacking/

[6] M. N. O. Sadiku, S. Alam, S. M. Musa, and C. M. Akujuobi, "A primer on cybersecurity," *International Journal of Advances in Scientific Research and Engineering*, vol. 3, no. 8, Sept. 2017, pp. 71-74.

[7] M. N. O. Sadiku, M. Tembely, and S. M. Musa, "Smart grid cybersecurity," *Journal of Multidisciplinary Engineering Science and Technology*, vol. 3, no. 9, September 2016, pp.5574-5576.

[8] "FCC Small Biz Cyber Planning Guide,"

https://transition.fcc.gov/cyber/cyberplanner.pdf

[9] "The 8 most common cybersecurity attacks to be aware of,"

https://edafio.com/blog/the-8-most-common-cybersecurity-attacks-to-be-aware-of/

[10] Y. Li and Q. Liu, "A comprehensive review study of cyber-attacks and cyber security; Emerging trends and recent developments," *Energy Reports*, vol. 7, November 2021,

https://www.sciencedirect.com/science/article/pii/S2352484721007289

[11] Y. Zhang, "Cybersecurity and reliability of electric power grids in an interdependent cyber-physical environment," *Doctoral Dissertation*, University of Toledo, 2015.

[12] "Construction cyber risk in the construction industry: An important consideration," August 2016,

https://esub.com/blog/construction-cyber-risk-in-the-construction-industry/

[13] F. Osborn, "Safeguarding your construction company from cyber threats," September 2024,

https://www.foundationsoft.com/learn/safeguarding-your-construction-company-from-cyber-threats/

[14] "Combatting cyber threats in the construction industry," February 2024,

https://www.captechu.edu/blog/combatting-cyber-threats-construction-industry

[15] "Why the construction industry is being impacted by cyberattacks, and what to do about it,"

https://www.agc.org/sites/default/files/Galleries/enviro_members_file/CLE%20Paper_%20Cyber%20Attacks%20and%20the%20Construction%20Industry.pdf

[16] D. Anderson, B. Q. Choi, and J. Valdez, "Building defenses

against cyber risk in the construction sector," March 2024, https://woodruffsawyer.com/insights/cybersecurity-in-construction

INDEX

www.ingramcontent.com/pod-product-compliance
Lightning Source LLC
LaVergne TN
LVHW070827220926
848486LV00027B/30
9798991502757